How Bad Are Bananas?

MIKE BERNERS-LEE

How Bad Are Bananas?

THE CARBON FOOTPRINT OF EVERYTHING

GREYSTONE BOOKS

D&M PUBLISHERS INC.

Vancouver/Toronto/Berkeley

Copyright © 2011 by Mike Berners-Lee
First published in Great Britain in 2010 by Profile Books Ltd.

11 12 13 14 15   5 4

All rights reserved. No part of this book may be reproduced,
stored in a retrieval system or transmitted, in any form
or by any means, without the prior written consent of the publisher
or a license from The Canadian Copyright Licensing Agency
(Access Copyright). For a copyright license, visit
www.accesscopyright.ca or call toll free to 1-800-893-5777.

Greystone Books
An imprint of D&M Publishers Inc.
2323 Quebec Street, Suite 201
Vancouver BC Canada V5T 4S7
www.greystonebooks.com

Cataloguing data available from Library and Archives Canada
ISBN 978-1-55365-831-3 (pbk.)
ISBN 978-1-55365-832-0 (ebook)

Cover and text design by Jessica Sullivan
Cover photograph by John Sherlock
Printed and bound in Canada by Friesens
Text printed on acid-free, 100% post-consumer paper
Distributed in the U.S. by Publishers Group West

FSC
www.fsc.org

MIX
Paper from
responsible sources
FSC® C016245

Contents

Under 10 grams

10 grams to 100 grams

100 grams to 1 kilo (2.2 pounds)

1 kilo to 10 kilos
(2.2 pounds to 22 pounds)

Introduction

A FEW YEARS AGO I agreed to go around a supermarket with a journalist who wanted to write an article on low-carbon food. We trailed up and down the aisles with the Dictaphone running, and she plied me with questions, most of which I was pitifully unable to answer.

"What about these bananas?... How about this cheese?... It's organic. That must be better... isn't it?... Or is it?... Lettuce must be harmless, right?... Should we have come here by bus?... At least we didn't fly! How big a deal is food, anyway?"

It was not at all clear what the carbon-conscious shopper should do. There was clearly a huge gap in the available consumer knowledge, and, on that day, we couldn't fill it. The article never happened, and it's probably just as well. Since then I have looked long and hard into all kinds of carbon footprints and carried out numerous studies, including one for a supermarket chain.

This book is here to answer that journalist's questions, and many more besides. It's not just a book about food and travel. I want to give you a sense of the carbon impact—that is, the climate change impact—of everything you do and think about. I want to give you a carbon instinct. Although I have discussed the footprint of just under one hundred items, I hope by the time you have read about these you will have gained such a sense of where carbon impacts come from that you will be able to make a reasonable guesstimate of the footprint of more or less anything and everything that you come across. It won't be exact, but I hope you'll at least be able to get the number of zeros right most of the time. There are messages in this book for ordinary people, some for businesses, and a few sprinkled in for policy makers too.

Some basic assumptions

I'm hoping I can take three things for granted:
> climate change is a big deal;
> it's caused by humans,
> and we can do something about it.

However, out of respect for the still-widespread confusion over these assumptions, I have put more about them in an appendix in case you want to check them out before moving on.

Perspective

A friend recently asked me how he should best dry his hands to reduce his carbon footprint—with a paper towel or with an electric hand drier. The same person flies across the Atlantic literally dozens of times a year. A sense of scale is required here. The flying is tens of thousands of times more important than the hand drying. So my friend was simply distracting himself from the issue. I want to help you get a feel for roughly how *much* carbon is at stake when you make simple choices—where you travel, how you get there, whether to buy something, whether to leave the TV on standby, and so on.

Picking battles

I'm not trying to give you a list of 500 things you can do to help save the planet.[1] You could probably already write that list yourself. You will find at least 500 possibilities in here, but this is a book about helping you work out where you can get the best return for your effort. This book is here to help you *pick your battles*. If you enjoy the read and by the end of it have thought of a few things that can improve your life while cutting a decent chunk out of your carbon, then I'll be happy. The book isn't here to tell you what to do or how radical to be. Those are personal decisions.

Is carbon like money?
In one sense, yes it is.

Carbon is just like money in that you can't manage it unless you understand it, at least in broad terms. Most of the time we know how much things cost without looking at the price tag. I don't mean that we have an exact picture, but we know that a bottle of champagne is more expensive than a cup of tea but a lot cheaper than a house. So most of us don't buy houses on a whim. Our financial sense of proportion allows us to make good choices. If I really want champagne, I know I can have it, provided that somewhere along the line I cut out something just as expensive that is less important to me. Our carbon instinct needs to be just like the one we have for managing our money.

That's where the similarity ends. Unlike with money, we are not used to thinking about carbon costs. It's also much harder to tell how much we are spending because we can't see it and it's not written down. Furthermore, unlike what happens when we spend a lot of money, we don't personally experience the consequences of our carbon impact because it's spread across nearly seven billion people and many years.

Enjoy the read
These pages are written for people who want to love their lives and for whom that now entails having some carbon awareness alongside everything else that matters to them.

Dip in. Keep this book by the loo. Read it from cover to cover or flit around. Use it as a reference if you like. Talk about it. Take issue with it. Let me know how it could be improved (info@howbadarebananas.com). Think of it like an early map, full of inaccuracies but better, I hope, than what you had before.

If there's something else to be gleaned from the book, it is that nearly all of us, including me, have plenty of junk in our lives that contributes nothing at all to the quality of our existence. It's deep in our culture. Cutting that out makes everyone's life better, especially our own. I got a big win by swapping my solo car commutes for bike rides and carpools. That works for me, but I'm not prescribing that particular solution for you, because we are all different. I hope you enjoy the read and that, while you are at it, you bump into at least something you can use.

So how bad *are* bananas?

As it happens, they turn out to be a fine low-carbon food, though not totally free from sustainability issues to keep an eye on—see page 31.

A quick guide to carbon and carbon footprints

Carbon footprint is a lovely phrase that is horribly abused.[1] I want to make my definition clear at the outset.

Throughout this book, I'm using the word *footprint* as a metaphor for the total impact that something has.

And I'm using the word *carbon* as shorthand for all the different global warming greenhouse gases.

So, I'm using the term *carbon footprint* as shorthand to mean the *best estimate* that we can get of the *full climate change impact* of something. That something could be anything—an activity, an item, a lifestyle, a company, a country, or even the whole world.

CO_2e? What's that?

Human-caused climate change, also known as global warming, is caused by the release of certain types of gas into the atmosphere. The dominant greenhouse gas generated by humans is carbon dioxide (CO_2), which is emitted whenever we burn fossil fuels in homes, factories, or power stations. But other greenhouse gases are also important. Methane (CH_4), for example, which is emitted mainly by agriculture and landfill sites, is 25 times more potent per pound than carbon dioxide. Even more potent but emitted in smaller quantities are nitrous oxide (N_2O), which is about 300 times more potent than carbon dioxide and released mainly from industrial processes and farming, and refrigerant gases, which are typically several thousand times more potent than carbon dioxide.

In the U.S., the total impact on the climate breaks down like this: carbon dioxide (85 percent), methane (8 percent), nitrous oxide (5 percent), and refrigerant gases (2 percent).[2]

Given that a single item or activity can cause multiple different greenhouse gases to be emitted, each in different quantities, a carbon footprint, if written out in full, could get pretty confusing. To avoid this, the convention is to express a carbon footprint in terms of *carbon dioxide equivalent* (CO_2e). This means the total climate change impact of all the greenhouse gases caused by an item or activity rolled into one and expressed in terms of the amount of carbon dioxide that would have the same impact.[3]

Beware carbon toe-prints

The most common abuse of the phrase *carbon footprint* is to miss out some or even most of the emissions caused, whatever activity or item is being discussed. For example, many online carbon calculator websites will tell you that your carbon footprint is a certain size based purely on your home energy and personal travel habits, while ignoring all of the goods and services you purchase. Similarly, a magazine publisher might claim to have measured its carbon footprint but in doing so looked only at its office and cars while ignoring the much greater emissions caused by the printing house that

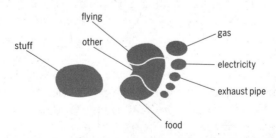

FIGURE 1.1: The footprint of a lifestyle is bigger than its toe-print.

produces the magazines themselves. These kinds of carbon footprints are actually more like carbon "toe-prints"—they don't give the full picture.

Direct and indirect emissions

Much of the confusion around footprints comes down to the distinction between "direct" and "indirect" emissions. The true carbon footprint of a plastic toy, for example, includes not only direct emissions resulting from the manufacturing process and the transportation of the toy to the store; it also includes a whole host of indirect emissions, such as those caused by the extraction and processing of the oil used to make the plastic in the first place. These are just a few of the processes involved. If you think about it, tracing back all the things that have to happen to make that toy leads to an infinite number of pathways, most of which are infinitesimally small. To make the point clearly, let's try following just one of those pathways. The staff in the offices of the plastic factory use paper clips made of steel. Within the footprint of that steel is a small allocation to take account of the maintenance of a digger in the iron mine that the steel originally came from ... and so on forever. The carbon footprint of the plastic toy includes the lot, so working it out accurately is clearly no easy task!

To give another example, the true carbon footprint of driving a car includes not only the emissions that come out of the tailpipe but also all the emissions that take place when oil is extracted, shipped, refined into fuel, and transported to the gas station, not to mention the substantial emissions caused by producing and maintaining the car.

A note about high-altitude emissions

Emissions from planes in the sky are known to have a greater impact than those that would arise from burning the same amount of fuel at ground level. The science of this discrepancy is still poorly understood. Nevertheless, because our measure is setting out to be a guide to climate change impact, it is essential to try to take this into account. That is why in this book I have multiplied all aviation emissions by 1.9.[4] (Some experts believe the true impact of plane emissions could be even higher and suggest a multiplier of up to 4.)

In the Further information section you'll find a slightly more technical discussion of the methodologies I have used to get the numbers (page 192).

The essential but impossible measure

The carbon footprint, as I have defined it, is *the* climate change metric that we need to be looking at. The problem is that it is also impossible to measure. We don't stand a hope of being able to understand how the impact of our bananas compares with the impact of all the other things we might buy instead unless we have some way of taking into account the farming, the transport, the storage, and the processes that feed into those stages. A key question, then, is this: How should we deal with a situation in which the thing we need to understand is impossibly complex?

One common response is to give up and measure something easier, even if that means losing most of what you are interested in off the radar. The illusionist Derren Brown refers to one of his core techniques as the *misdirection of attention:* by focusing his audience on something irrelevant, he can make them miss the bit that matters. Examples include an airport waxing lyrical about the energy efficiency of its buildings without mentioning the flights themselves. The same thing can happen by accident. If you settle for a toe-print, there is a very good chance it will misdirect your attention and steer it away from the big deals.

An alternative response to the problem, and the approach that this book is all about, is to do the best job you can, despite the difficulties, of understanding the whole picture. This book is about making the most realistic estimates that are possible and practical and being honest about the uncertainty.

Blurry numbers...

First and foremost, I am trying to get the orders of magnitude clear.

In my work I put a lot of effort into developing a realistic picture of different carbon footprints using a variety of methods. This book draws upon a lot of that, as well as the most credible secondary sources that I have been able to find. However, huge uncertainty remains. So when you see a number like "2.5 kg (5.5 lbs.) CO_2e" on an item such as a burger, bear in mind that it is a best estimate. What it really means is something like "best estimate of 2.5 kg (5.5 lbs.) CO_2e, probably between 1.5 and 4 kg (3 and 8 lbs.) CO_2e and almost certainly between 1 and 10 kg (2 and 20 lbs.)." That is the nature of all carbon footprints. Don't let anyone tell you otherwise.

Some of the numbers you'll see are even flakier still. This generally happens when I'm trying to bring the beginnings of a sense of scale to important questions that are almost impossible to quantify. Sometimes my calculations and assumptions are highly debatable, but I've included them because I think that just going through the thought process can be a useful reflection on something that matters. Examples include the footprint of having a child, waging a nuclear war, or sending a text message.

If you think you can offer an improvement on any of the numbers in this book, I'll be very happy to hear from you (info@howbadarebananas.com).

... but they will do...

Let me be emphatic that the uncertainty does not negate the exercise. Real footprints are *the essential measure,* and nothing short of them will do. The level of accuracy that I have described is good enough to separate out the flying from the hand drying. And even if you use the numbers here to make finely balanced decisions, most of the time those choices will be better informed than if you had no guidelines at all.

... for now

That we find footprinting tricky is a problem for us all. The situation we are in is like sailing round the world with a map from the 1700s. How should we respond? Throw that map away and have nothing? Definitely not! Use a high-quality map of just a small part of the ocean and ignore the rest? No way. Use the maps we have but treat them with caution? Absolutely. Try to make better maps? Of course—and the work is ongoing. This book is just an early map. Better ones will follow. And this book is trying to help you improve the carbon map that you carry around in your own head.

Making sense of the numbers

So far we've established what we need to try to measure, but a ton of carbon is still a highly abstract concept. I'm now going to try to give it a bit more real-life meaning.

A note about units

When this book was first published in the U.K., the measurements were given in grams, kilograms, and tonnes (metric tons). For this North American edition, I have continued to use grams, kilos and tons of CO_2e as my units for carbon footprints, since that allows us to use a decimal scale that allows a straightforward comparison of impacts. However, I have added some conversions for clarity—in particular, I have given the pound equivalents for measurements in kilograms. So, for example, the carbon footprint of asparagus is described in kg CO_2e per pound.

I have not, however, offered any conversions to most of the measurements in grams. One gram is the mass of about 15 drops of water, and there are 30 grams in an ounce. Because there is such a large disparity between a gram and an ounce, including conversions would lead to unwieldy strings of numbers with a lot of decimal places that detract from the bigger picture.

At the other end of the spectrum are tons, and since a tonne, or metric ton, is reasonably close to a U.S. ton, I've simply given measurements in tons only, without bothering with any conversions. I took the same approach with liters and quarts, which are fairly close, and all gallons mentioned in the text are U.S. gallons.

What does a ton of CO_2e look like?
If you filled a couple of standard-sized 60-gallon garden water tanks to the brim with gasoline and set fire to them, about a ton of carbon would be directly released into the atmosphere. (The carbon footprint of burning that gas by driving is a bit more than that, for reasons explained later.) If you did the same with two cups of gas, that would release just over a kilogram (2 pounds) of carbon dioxide, and if you burned a blob about the size of a chickpea, that would release about a gram. There are a thousand grams in a kilogram and roughly a thousand kilograms in a ton.

How many tons do we each cause?
To give a quick sense of scale, the average North American currently has an annual carbon footprint of around 28 tons. Australians are a tad higher still at around 30 tons, while for the U.K. the figure is around 15 tons. The Chinese weigh in at just 3 tons per head and the Malawians at a tiny 100 kg (about 220 lbs.). Our global average is about 7 tons each. There is more detail on this later on. You get smaller numbers if you include only the obvious bits of your footprint such as household energy and travel or you miss out emissions on goods you buy that are manufactured overseas.

The 10-ton lifestyle
I'm not here to set you a particular target or to make you feel guilty. How you decide to live is a choice that only you can make. I just want to help you understand carbon so that you can do whatever you decide to do with more knowledge.

However, to help get a sense of perspective, I have adopted a 10-ton lifestyle as another unit of measure for this book. I am going to refer to it from time to time, because it gives an alternative and sometimes clearer way of conceiving of those abstract grams and tons of CO_2e.

There is not much that is particularly magical about a 10-ton lifestyle—that is, a lifestyle causing 10 tons of CO_2e per year—apart from the fact that 10 tons is a round number. It's certainly not a long-term sustainable target for everyone in the world: if everyone went in for 10-ton living all over the globe, emissions would skyrocket by 40 percent.

I've chosen 10 tons as a realistic target that most people ought to be able to meet. For the average North American it would mean a reduction of about 65 percent. In the U.K. and many other European countries, it entails reducing your emissions to only about one-third below your national average.

One way of thinking about the footprint of an object or activity is to put it in the context of a year's worth of 10-ton living. For example, a large cheeseburger, with a footprint of 2.5 kg (5.5 lbs.) CO_2e, represents about 2 hours' worth of a 10-ton year. If you drive a fairly thirsty car for 1,000 miles, that is 800 kg (1,750 lbs.) CO_2e, or a month's ration. If you leave a couple of the (now old-fashioned) 100-watt incandescent light bulbs on for a year, that would be another month used up. One typical return flight from Los Angeles to Barcelona burns up around 4.6 tons CO_2e. That is just under 6 months' ration in the 10-ton lifestyle.

Truly sustainable long-term targets aren't practical or helpful in the short term. For example, the U.K.'s target to cut carbon emissions by 80 percent by 2050 is currently the most ambitious of any major economy. If you were to apply this to the stuff they import as well as to the emissions within the country itself, that would take them down to around 3 tons per person per year. Some commentators think they'll need to go even lower. Ultimately, though, it's virtually impossible for an individual in the developed world to get down to a 3-ton lifestyle anytime soon. That kind of cut requires the whole economy to be made greener. More modestly, the government of Canada is committed to reducing total greenhouse gas emissions by 17 percent from 2005 levels by 2020, aligned with U.S. targets.[5]

A short car commute, a daily cheeseburger, and some wasteful lighting habits could easily use up a quarter of the 10-ton budget. Then if you also take the flight from L.A. to Barcelona, that would leave just 3 months' ration left in the 10-ton budget for *everything* else that year: other food, heat, buying stuff, health care, use of other public services, your contribution to the maintenance of roads, any wars around the world that your government is involved in (like it or not)— *the lot*.

You might be wondering whether there are any better ways of spending this or any other sized budget besides blowing most of it on burgers, commuting, and flying. If that question is of interest, this book should give some clues.

How many tons for a life or a death?

I hope the comparisons so far have helped to make a ton of carbon seem a bit more tangible. But let's see whether it's possible to get a handle on how much it might actually *matter*. Our species is good at understanding the direct, immediate, and visible consequences of our actions. We are a lot less smart at grasping the consequences when they are dispersed across billions of people whom we will never meet. This might not have mattered when we lived in caves, but it won't let us live well in a global society. Our impacts used to be local and visible. Today they are not. Perhaps we need to find it as shocking when we see dispersed suffering inflicted through needless carbon emissions as it would be to see the same suffering inflicted all in one place in front of our eyes by, let's say, a street stabbing.

I did some "back of the envelope" sums and arrived at a figure of 150 tons CO_2e per climate change–related death. I've spelled out my calculations in the endnote that follows this sentence.[6] If you look it up and follow my sums, you'll see that I don't have even the beginnings of a rigorous argument to justify my figure. But it was an interesting thought process and one that, if you do decide to follow it, you might even find faintly plausible. Or you may think my line of thought is hopelessly unrealistic. And maybe you would be right. I was just playing with ideas. It is up to you to decide what meaning to take from them. For me, even a possibility of any realism in this line of thought throws up a challenge.

The 150 tons per life figure would mean that if your lifestyle had the footprint of the average U.S. citizen, one person would have to die from climate change somewhere in the world every 5 years. If you were to fly from L.A. to Barcelona and back 11 times first class—that would be another death.

How much would it be worth paying to save a ton of carbon?

This is not going to be an easy question to answer. An unknown number of lives depends on our response to climate change, and even if we did know how many, it is not as if our society has a consistent approach, even in the very broadest of terms, to determine the kind of value that each one of those lives might have. So, putting a financial value on saving a ton of carbon is going to be tough, to put it mildly. Nevertheless, it's a question worth pondering because unless we understand there to be real and tangible value in cutting emissions, we will simply never bother and, for better or worse, money has become our language for understanding value.

As I write, the equivalent of $18 per ton is the maximum price of CO_2 that companies in the U.K. could have to pay.[7] Let's see what happens if we work on that figure. With global emissions at 50 billion tons, does that mean that the world might be prepared to pay just $900 billion to eradicate our emissions completely? Is that really all it's worth to us? That's about three-quarters of a percent of global output in economic terms to have a miracle cure for climate change? A bargain, surely.

Let's see what $18 per ton implies if you link it in to my estimate of 150 tons per death. That would put the value of a life at just $2,700. The value of the world's population under this analysis is a mere $18 trillion, or about six times the Gross Domestic Product of the U.K. My hometown of Kendal has about 24,000 people. Would it really be a good deal to blow up everyone in it if it would liberate $65 million? This analysis places the value of the U.K. population at just $164 billion. In other words, the people living in the U.K. are valued at about 5 percent of their GDP.

So how much should it be worth in financial terms to save a ton of carbon? A great deal more than the $18, clearly!

Under 10 grams

A text message

0.014 g CO₂e one message
32,000 tons CO₂e all world's texts for a year

The biggest part of a text message's footprint is the power used by your phone while you type—and of course by your friend's phone while he or she reads what you've written. If the two of you take a minute between you to type and read the message, and you each have phones that consume 1 watt of power when in use, the message's footprint will be about a hundredth of a gram. This figure takes into account the transmission of a 140-character message across the network.[1]

Around the world, about 2.5 trillion texts are sent every year.[2] Don't be fooled into thinking that the 32,000-ton footprint for this total is a big number. It isn't. 32,000 tons is about one ten-thousandth of a percent of the world's carbon footprint. In other words, texting is not a big deal. It wouldn't even be a big deal if my numbers were out by a factor of a hundred.

Incidentally, as of 2008, nearly a quarter of all text messages were sent in China, and about a fifth in the Philippines, where they average an impressive 15 messages per day for each phone. The average North American phone sent just a couple of messages a day, whereas British phones manage six texts per handset.

In summary, we can relax about sending texts (but no spam, please).

A cup of tap water

0.06 g CO_2e one cup
23 kg (51 lbs.) CO_2e a year's tap water for a typical U.S. citizen
> A year's supply for one person is the same as a 27-mile drive
in an average passenger car.[3] That includes drinking, washing,
cleaning—the lot.

Unlike the bottled alternative, which has around 1,000 times
the impact (see page 44), cold tap water is not a major carbon
concern for most North Americans. In the U.S., for example,
the provision and disposal of household water accounts for
less than third of a percent of the national carbon footprint.[4]
Climate change looks set to cause serious water stress in some
places while other areas are going to have plenty. Interestingly, if our cup of tap water is poured down the drain, its
footprint leaps almost fourfold to a quarter of a gram because
it is more carbon intensive to treat waste water than to supply
the water in the first place.[5] If the eventual fate of the drink is
to be flushed down the loo along with another couple of gallons, that takes the total to 4 g CO_2e.

Tap water itself is one thing. Heating it up is another matter, accounting for a decent chunk of the typical person's
emissions (see page 137). See also Swimming pool, page 150,
and Desalination, page 93.

A web search

0.2 g CO_2e Google's estimate for the energy used at their end
0.7 g CO_2e from an efficient laptop (a lower estimate)
4.5 g CO_2e from a power-hungry machine and making higher
estimates of power used in the network
> So that is between 2 and 14 seconds' worth of 10-ton living for
a 30-second single search.

At the low end of the scale, I've started off with Google's estimate of 0.2 g CO_2e for the electricity they use at their end when
you put in a single search enquiry.[6] Add to it just 30 seconds

of machine time at your end on an efficient 20-watt laptop while you tap in the search, wait for the result, and scan it for what you want. That's another 0.1 g, bringing the total so far to 0.3 g. Your local network and the servers that actually host the information you are digging for probably come to at least 50 percent of the amount of power used by your machine, even if they are super-efficient, like your laptop,[7] so that takes us to 0.35 g. Wear and tear and depreciation of hardware throughout the whole system probably doubles this because of the emissions that are required in the manufacture of all that kit. That takes us to 0.7 g CO_2e for a single enquiry that might let you, say, find the location of the restaurant you're heading to.

On a more power-hungry desktop computer that uses 150 watts of power, your web search might burn through about 0.75 g CO_2e. If you apply the same markups for networks and hardware, we get to a grand total of 4.5 g, with Google accounting for just 0.2 g of that.

You can search for information about the footprint of web searches. You'd find blogs and articles all coming up with different figures based on different assumptions and all including different things. Some look at multiple searches and therefore produce much higher headline figures.[8]

At the high end of my estimate, the activity of surfing clocks up a carbon footprint at about half the rate of the 10-ton life. In other words, if you spent a whole year browsing the web nonstop, you'd trigger about 5 tons of emissions. That sounds good until you remember that at the same time you might also be wearing clothes, keeping warm, burning calories, getting closer to your next need for medical attention, living in a building that needs periodic maintenance, and so on. Even while you are sitting at the machine, your browsing is just one part of your footprint.

Google is estimated to deal with 200 to 500 million enquiries per day. If we go with the top estimate, and the high-end figure for the footprint of a single search, Google searching accounts for 1.3 million tons CO_2e per year. That is a big number, but it is only about one forty-thousandth of our global footprint. We can probably relax about it. Reading the stuff we find is an altogether more carbon-hungry activity—see page 123.

Walking through a door

Zero CO_2e a normal household door on a summer's day
3 g CO_2e getting in through your front door on a cold New York winter's day
84 g CO_2e big electric doors opening into a large stairwell on a cold, windy day in New York

> At the high end, that's a banana's worth of greenhouse gas every time you enter the building.

The entrance door of the building where I work has no manual option.[9] To get in, you have to press a button and wait while two electric motors whir and double doors swing slowly open, creating a space 2 m wide by 2.5 m high (about 6.5 feet wide by 8 feet high). You enter a spacious stairwell with two large radiators. It takes 18 seconds for the doors to finish closing. This three-year-old building was amazingly rated environmentally "Excellent."[10]

The power used by the electric motors themselves isn't the problem. They account for just 1 g CO_2e. The problem is the size of the space you have to open, the time it has to stay open for, and the vast heated space that the doors open onto.

For this building there must have been lots of other options, such as manual doors that swing shut and can be opened singly, with an override button for disabled access. Rotating doors attached to turbines that generate electricity as you pass through have been trialed in Holland but sound like the kind of gimmick that can tarnish the reputation of the renewables industry.

In a typical home on a cold, blustery day, the numbers are more likely to come out at about 3 g, based on opening it by hand and closing it straight away.

An email

0.3 g CO_2e a spam email
4 g CO_2e a proper email
50 g CO_2e an email with long and tiresome attachment that you have to read

> A typical year of incoming mail adds up to 135 kg (300 lbs.) CO_2e: over 1 percent of the 10-ton lifestyle and equivalent to driving 200 miles in an average car.

The annual figure provided here is for the typical business user and includes the sending, filtering and reading of every incoming message. According to research by the spam and virus protection specialists, McAfee, a remarkable 78 percent of those incoming emails are spam. Around 62 trillion spam messages are sent every year, requiring the use of 33 billion units of electricity and causing around 20 million tons of CO_2e per year. McAfee estimated that around 80 percent of this electricity is consumed by the reading and deleting of spam and the searching through spam folders to dig out genuine emails that ended up there by accident. Spam filters account for 16 percent. The actual generation and sending of the spam is a very small proportion of the footprint.

Although 78 percent of incoming emails sent are spam, these messages account for just 22 percent of the total footprint of your email account, because, although they are a pain, you deal with them quickly. Most of them you never even see. A genuine email has a bigger carbon footprint, simply because it takes time to deal with. So if you are someone who needlessly copies people in on messages just to cover your own back, so that you can claim they should have known about it, the carbon footprint gives you one more good reason for changing your ways. You may find that after a while everyone at work starts to like you more, too.

The average email has just one-sixtieth the footprint of a letter (see page 45). That looks like a carbon savings unless you end up sending 60 times more emails than the number of letters you would have mailed in days gone by. Lots of people do. This is a good example of the *rebound effect*—a low-carbon technology resulting in higher-carbon living simply because we use it more.

If the great quest is for ways in which we can improve our lives while cutting carbon, surely spam and unnecessary email have to be very high on the hit list, along with old-fashioned paper junk mail.

If only email were taxed. Just a cent per message would surely kill all spam instantly. The funds could go to tackling world poverty, say. The world's carbon footprint would go down by 20 million tons even if genuine users didn't change their habits at all. The average user would be saved a couple of minutes of their time every day, and there would be a $170 billion annual fund made available. If one cent turned out to be enough to push us into a more disciplined email culture—with perhaps half the emails sent—the anti-poverty fund would be cut in half, but a good few minutes per day would be liberated in many people's lives, and the carbon savings would be around 70 million tons CO_2e—that's nearly 10 percent of all of Canada's emissions.

Drying your hands

Zero CO_2e letting them drip
3 g CO_2e Dyson Airblade
10 g CO_2e one paper towel
20 g CO_2e standard electric drier
> On average, if you used public toilets six times per day, your hand drying would produce around 15 kg (33 lbs.) CO_2e per year; equivalent to 1 kg (2.2 lbs.) of beef.

"What's the greenest way to dry my hands?" is a frequently asked question, so I'll answer it even though I have already made the point that if you really want a lower-carbon lifestyle, you should be asking about something more important.

Close to the low end of the scale is drying your hands with a Dyson Airblade. This dryer does the job in about 10 seconds with 1.6 kilowatts of power. Its secret is that it doesn't heat the air. It just blows it hard. This makes it far more efficient than conventional hand driers.

In the middle of the spectrum I have put paper towels, based on 10 g of low-quality recycled paper per sheet, and only one towel used each time.[11] (Of course, if you use two or three towels, the footprint doubles or triples.)

At the high end are conventional heated hand driers. These take a shade longer than the Dyson and use around 6 kilowatts of power. The big difference is explained by the fact that it always takes a lot of energy to create heat.

Right at the bottom of the scale comes not drying your hands at all—or indeed using a small hand towel that is reused many times in between low-temperature washes. I am not a hygiene expert, but I'm told that neither option is good from that point of view: they may even end up adding to the already substantial footprint of the health service (see page 129).

A plastic carrier bag

3 g CO_2e very lightweight variety
10 g CO_2e standard disposable supermarket bag[12]
50 g CO_2e heavyweight, reusable variety
> So that's 2.5 kg (5.5 lbs.) per year if you use five standard bags per week: about the same as one large cheeseburger.

Over the past few years many supermarkets have been making an effort to reduce the use of plastic bags, and Rwanda has set a dazzling example for the world by banning them altogether. This is fantastic news for other environmental reasons but doesn't constitute a response to climate change. When someone in the developed world walks home from the store with a disposable plastic bag full of food, the bag is typically responsible for about one-thousandth of the footprint of the food it contains. In other words, it is good if your supermarket is taking action on plastic bags, but don't let that stop you from asking what it is doing about the other 999 thousandths of its carbon agenda.

Carbon emissions are not the only environmental problem associated with plastic bags, of course. They also have a habit of hanging around in the ecosystem where they can sit for hundreds of years, clogging up animals' stomachs, killing fish, and being ugly. *National Geographic* estimates that the world uses between 500 billion and 1 trillion disposable grocery bags per year.[13] That's an awful lot of trash—even if the bags contribute only around one ten-thousandth of the world's total carbon footprint.

How best to get rid of them, then? Burning releases nasty toxins as well as carbon, although the technology is improving. From a purely climate change perspective, landfill is not too bad. They won't degrade, so all those hydrocarbons are returned to the ground where they came from for fairly long-term storage. But landfill is nasty for other reasons.

So, although disposable plastic bags aren't a serious carbon issue, they are still nasty for other reasons. Better alternatives are a backpack (which makes things easier to carry and keeps your hands free), a wheelie basket (which prevents you from having to lift things at all), or sturdy, reusable bags. If you do use reusable plastic bags, make sure you really do reuse them: if you get less than five uses out of one, you'd be better off, in carbon terms, with disposable ones.

10 grams to 100 grams

A paper carrier bag

12 g CO$_2$e recycled and lightweight
80 g CO$_2$e an elaborate bag from mainly virgin paper as supplied by many clothing retailers

A common misconception is that paper bags must be lower carbon than plastic. Wrong! The paper industry is highly energy intensive. Printed virgin paper typically produces between 2.5 and 3 kg CO$_2$e per kilo (between 1.1 and 1.4 kg CO$_2$e per pound) of paper manufactured. This is comparable with the emissions required to produce 1 kg (2.2 lbs.) of polypropylene plastic bags. However, paper bags have to be much heavier, so overall the paper bag ends up having a bigger footprint.

Recycled paper is roughly half as energy intensive to produce as virgin paper. But even a lightweight recycled paper bag produces slightly more greenhouse gas emissions than a typical plastic carrier.

There is another problem at the disposal end as well, which I have not factored into my numbers. Unless you recycle your paper bag, it is likely to end up in landfill, where it will rot, emitting more CO$_2$ and, even worse, methane. Landfill sites vary in their ability to capture and burn methane emissions, but typically there will be around 500 g of greenhouse gas emissions per kilo (or around 1 lb. of gas emissions per 2 lbs.) of paper buried.[1]

One final detail about paper bags is that they often don't work, resulting in bruised apples rolling down the street.

Low-carbon tips
> If given a choice between plastic and paper, the plastic one may well be best (see page 22).
> If stuck with paper, recycle it when you are done with it. (It is probably too much to hope that it could be fit for reuse.)

Ironing a shirt

14 g CO_2e a quick, expert skim on a slightly damp shirt
25 g CO_2e average
70 g CO_2e a thoroughly crumpled shirt ironed by unskilled hands
> Five shirts every week is about the same as a 10-mile drive once a year in an average car.

A friend of mine used to iron her husband's socks (she's now divorced). If you're feeling stuck in a similar routine, I hope you will find the carbon argument gives a bit more power to your elbow.

Although ironing isn't the biggest environmental issue, there may be scope for saving a little bit of carbon here—and perhaps some lifestyle improvement, too. For ironing that simply has to be done, the best green step is to have the clothes slightly damp and use the ironing process itself to finish off the drying. That saves both time and carbon (especially if you otherwise would be using an energy-hungry tumble drier—see page 86). Even more effective is simply using the iron less often.

A few people allegedly enjoy this activity, almost as a hobby. If ironing is how you get your kicks, it works out at about 400 g CO_2e per hour. That's about five times worse than watching the average TV but dramatically better than going for a drive. I have also heard ironing described as having meditative value. I can only assume that this goes something along the lines of "a deep reflection on the resentment you notice inside yourself at spending your time in this way." If this is you, can I recommend a good old-fashioned, Zen-style breathing routine, weighing in at zero g CO_2e?

Cycling a mile

65 g CO_2e powered by bananas
90 g CO_2e powered by cereals with milk
200 g CO_2e powered by bacon
260 g CO_2e powered by cheeseburgers
2,800 g CO_2e powered by air-freighted asparagus
> If your cycling calories come from cheeseburgers, the emissions per mile are about the same as two people driving an efficient car.

I have based all my calculations on the assumption that you burn 50 calories per mile.[2] The exact figure depends on how fit you are (the fitter you are, the lower the figure), how heavy you are, how fast you go (the faster, the higher), and how much you have to use the brakes.

All that energy has to come from the food you eat, and that in turn has a carbon footprint. The good news is that the lower-carbon options are also the ones that make the best cycling fuel.

Bananas, of course, are brilliant (see page 31). Breakfast cereal is pretty good (let down slightly by the milk). The bacon comes in at around 200 g CO_2e for a 25 g slice with only enough calories for a mile and a quarter of riding.

As mentioned above, two people cycling along using calories from cheeseburgers would between them have about the same footprint as they would have had if they had shared a ride in an efficient car. At the ridiculous high end of the scale, however, is getting your cycling energy by piling up your plate with asparagus that has been flown by air from the other side of the world. At 2.8 kg (6 lbs.) per mile this is like driving a car that does just over 5 miles to the gallon. You'd be better off in a Hummer.

All my figures include 50 g per mile to take into account the emissions that are embedded in the bike itself and all the equipment that is required to ride it safely.[3] In the lower-carbon scenarios, the food accounts for only a small part of your impact, and the maintenance of bike and sundry equipment dominates.

Is cycling a carbon-friendly thing to do? Emphatically yes! Powered by biscuits, bananas, or breakfast cereal, the bike is nearly 10 times more carbon efficient than the most efficient of gas-powered cars. Cycling also keeps you healthy, provided you don't end up under a bus. (Strictly speaking, dying could be classed as a carbon-friendly thing to do, but needing an operation couldn't: see page 129).

Buying a folding bike so that I could commute on the train has been one of the best decisions I have made in recent years—in terms of both lifestyle and carbon. My journey takes 10 minutes longer, but I get half an hour's exercise and 15 minutes of reading a book each way. Because both of those are things I like doing but struggle to find enough time for, I've magicked an extra hour of the stuff I love into my day—while saving money and carbon.

One other thing: by taking my car off the road in rush hour, I cut everyone else's queuing time as well and reduce the emissions they belch out while they wait (see Congested car commute, page 107).

Boiling a quart of water

50 g CO_2e gas stove-top kettle, fairly low heat
70 g CO_2e electric kettle
115 g CO_2e saucepan on the gas without a lid and flames up the side

Some friends of ours have a stove-top kettle that they use on their gas stove, and we ended up debating the environmental pros and cons for months. Finally I spent half a morning measuring different methods. (A sad way of spending time, I know, but I did have a book to write.)

Our plug-in electric kettle was the fastest. Only 10 percent of the electrical energy was wasted, so although inefficiencies in our power stations and distribution systems make electricity a high-carbon way of producing heat, the electric kettle is still a fairly good way of boiling water at home.

How the gas kettle compares with the electric kettle depends on the time of year. In winter, our friends win the low-carbon prize easily. That's because although some of the heat from the gas flames escapes around the edge of their kettle, that heat isn't actually wasted: the kitchen is the heart of their house, so all the heat that goes into the room is useful. In their house, in fact, the gas stove is the most efficient form of heating because nothing is wasted up the flue (as it is with a gas furnace), nor is any heat sent to unoccupied rooms or lost in pipework (as it is with central heating).

In the summer, our friends still win the low-carbon prize provided they are willing to put their kettle on a small gas ring to maximize the proportion of the heat that goes into the water, rather than being lost around the sides. Doing this gives them a 30 percent carbon savings over the electric kettle but also means it takes three times as long (12 minutes) to boil. If they use large gas ring, the result is slightly *more* carbon than the electric kettle—and it's still 50 percent slower.

Saucepans turned out to be less efficient than stove kettles. It only makes sense to bring water to the boil in a saucepan if you are putting vegetables in at the start, in which case there is the benefit that they begin cooking a bit even before the water boils. If you do use a saucepan, keep the lid on (20 percent waste if you don't) and make sure the flames don't go up the sides (potential for another 20 percent waste).

To summarize, kettles are better than saucepans, and gas beats electric—but only if you are not in a hurry or you want to heat your room anyway. Just as important, of course, is not to boil more water than you actually need.

Four kettle design improvements are worth a mention, since there are some incredibly simple features waiting to hit the mass market.

> Although only about 10 percent of the heat generated by an electric kettle is wasted, I was surprised at how hard it was to find a kettle with proper insulation. Better insulation would also mean that if you forget it has boiled, or you accidentally boil more than you need, it would stay hot for longer.

> A kettle with a thermostat so that you can set it to 85°C (185°F) when that is all you need—such as when making coffee or herbal tea—is quicker, cheaper, and lower carbon, and it probably reduces the chance of mouth cancer. The Morphy Richards Ecolectric Kettle, so far not easily available in North America, is the only one I've found with this feature.

> An old-fashioned whistle or a beep option would stop you from forgetting when it has boiled.

> The Eco Kettle, already on the market, allows you to decant just the amount you need from a reservoir, making it easier to boil only what you need.

The Dragons' Den must be waiting for someone to put all these features together.

An apple

Zero CO_2e plucked from the garden
10 g CO_2e local and seasonal
80 g CO_2e average; that's 550 g per kilo (225 g per pound)
150 g CO_2e shipped, cold stored, and inefficiently produced
> Apples are a low-carbon food wherever they come from. Beyond that it is difficult to be certain about the details.

As you'd expect, local, in-season apples are best, but there is nothing particularly bad about buying them from anywhere in the world, as long as they travel on a boat rather than a plane. Indeed, in early summer, when any local apples will have been in cold storage for months, importing may be the lower-carbon option.

One study from a university in New Zealand found that apples grown in that country for the U.K. market incurred just 185 g CO_2e per kilo (90 g per pound)—significantly lower than U.K. apples for local consumption, which came in at 271 g per kilo (130 g per pound).[4] The argument made in the study was that U.K. production entailed greater use of fossil fuels on the farm and required more cold storage. The study also cited New Zealand's cleaner electricity mix. These factors, it claimed, outweighed the emissions from shipping the produce halfway around the world.

The same arguments can be applied to imports to parts of North America. Electricity in the U.S. is slightly more carbon intensive than in the U.K., but Canada's is on a par with that of New Zealand (see Unit of electricity page 56). The distances clearly vary depending on where you are shipping to.

A similar comparative study referenced by the U.K. government's Department for Environment, Food and Rural Affairs (Defra) produced similar orders of magnitude but found, conversely, that for Germany (which should be similar to the U.K.) local apples were more carbon friendly than those sourced from New Zealand.[5] It's difficult to unpick the arguments and determine who got closer to the truth. Each study went about things slightly differently and made different assumptions. This story illustrates an important point: these kinds of study are always tricky, heaped with far more uncertainties and subjective judgments than many people like to admit.[6]

One last point: as with all fruit and vegetables, it's a good idea to buy the most misshapen ones you can get, because that encourages the supply chain not to chuck them in the garbage before they ever reach the store.

A banana

Zero g CO_2e grown in your own garden
80 g CO_2e imported from the other side of the world
(or 480 g per kilo/240 g per pound)

> To answer the question in the title of this book, bananas aren't bad at all. They're brilliant! To emphasize the point, I'm eating one as I write.

Bananas are a great food for anyone who cares about their carbon footprint. For just 80 g of carbon, you get a whole lot of nutrition: 140 calories as well as stacks of vitamin C, vitamin B6, potassium and dietary fiber. Overall, they are a fantastic component of the low-carbon diet. Bananas are good for just about everyone—athletes, people with high blood pressure, everyday cycle commuters in search of an energy top-up, or anyone wishing to chalk up their recommended five servings of fruit and vegetables per day. There are three main reasons that bananas have such low carbon footprints compared with the nourishment they provide:

> They are grown in natural sunlight—no hot-housing required.
> They keep well, so although they are often grown thousands of miles from the end consumer, they are transported by boats (about 1 percent as bad as flying).
> There is hardly any packaging, if any, because they provide their own.[7]

On top of their good carbon and healthy eating credentials, the fair-trade version is readily available.[8]

Don't let me leave you with the impression that bananas, for all their good qualities, are too good to be true. There are environmental issues. Of the 300 types in existence, almost all those we eat are of the single, cloned "Cavendish" variety. The adoption of this monoculture in pursuit of maximum, cheapest yields has been criticized for degrading the land and requiring the liberal use of pesticide and fungicide. Furthermore, although land is dramatically better used for bananas than beef in terms of nutrition per acre, there are still parts of the world in which forests are being cleared for banana plantations[9] (see Deforestation, page 151).

Overall, however, the only really bad bananas are any that you let rot in your fruit bowl. These join the scandalous 40 to 50 percent of food wasted in the U.S.[10] and many other countries. If you do find yourself with bananas on the turn, they are good in cakes and smoothies. I have a distant childhood memory that they are also tasty in custard.

An orange

Zero g CO_2e grown in your own garden
90 g or 500 g CO_2e per kilo (230 g per pound) shipped 2,000 miles by boat and 500 by truck.
1 kg CO_2e each or 5.5 kg per kilo (2.5 kg per pound) air-freighted for the start of a season

Most oranges, along with most apples and bananas, are great from a carbon perspective.[11] They keep well and so can be grown in natural conditions and shipped around the world to wherever they are required.

The important thing to note here is that although there are often lots of food miles, these ones are usually fairly climate friendly. Like bananas, oranges can go on a huge boat and take their time. However, I was told by someone who buys fruit commercially that some supermarkets airfreight some varieties of orange at the start of the season to get them into the stores a couple of weeks early.

A quart of orange juice can have a footprint equivalent to several pounds of oranges. That's because orange juice incurs several inefficiencies in its production:
> The pulp is thrown out (so pulpy varieties and smoothies may be more sustainable).
> There are emissions from processing, including pasteurizing and sometimes turning into concentrate for transport purposes, and refrigeration.

> There is the footprint of the carton.
> Transport miles are often higher as the product moves from farm to juicer to cartoner to distributor, sometimes zigzagging wildly around the world.
> Fresh orange juice requires refrigeration. In the U.K. Tesco reports that its freshly squeezed juice has about twice the footprint of the long-life product. Most of that difference will be down to refrigeration.

An hour's TV

37 g CO₂e 15-inch LCD flat screen
84 g CO₂e 28-inch CRT TV
97 g CO₂e 32-inch LCD flat screen
240 g CO₂e 42-inch plasma screen

> One hour per day on the 32-inch LCD comes to 35 kg (77 pounds) CO₂e per year—equivalent to a 39-mile drive in an average gas-powered car.

Overall, watching TV turns out to be a remarkably low-carbon hobby, and it beats anything that involves driving. This is good news because the average American spends a massive four hours a day in front of the box (compared with just three and half hours for a European and a mere three hours for a typical Canadian).[12] You probably don't because you read books and, surely, there isn't time for both.

At its very worst, the 42-inch plasma screen, on for 10 hours per day, could clock up 880 kg (1,940 lbs.) CO₂e per year,[13] the equivalent of driving an average gas-powered car for about 940 miles. That may sound like a lot, but it actually makes for quite a low-carbon life, because it leaves so little time in your day to do anything else that might have a higher footprint.

The figures above don't take account of the emissions embodied in the TV set itself. The significance of these emissions—relative to the power the TV actually consumes in use—depends on what TV you have and how often you use it. The figure of 240 kg (530 lbs.) of CO_2e is a ballpark figure for the manufacture of a brand new TV costing $750, which at the time of writing is about the price of the energy-hungry 42-inch plasma version. This works out as 22 kg (49 lbs.) per year if you keep it for 10 years. If you watched that TV for 1 hour per day, the emissions from the electricity it will use, at about 80 kg (170 pounds) per year, will still dwarf those of the manufacture. At the other end of the scale, if you spend $300 on a 15-inch LCD, make it last just 5 years and watch it for only half an hour a day, the embodied emissions will dominate your TV footprint.

By watching with friends, you can clearly make things more efficient. The more people you invite around the better, provided they live within walking or cycling distance.

Should you replace your TV?

At my local waste disposal center (that's the place that used to be called the dump in the days before segregation) they currently have a whole room especially for homeless old-fashioned CRT televisions, most of which work fine but which are being disposed of to make way for modern flat-screen models. The people who run the disposal center say that, at the peak, they were taking in 400 CRT TVs per week.

So what are the carbon implications of trading in an old television for a new one? Figure 3.1 provides some answers. All the sums are based on these assumptions that your old TV is a typical 28-inch CRT model; that whatever choice you make now, you will stick with it for 10 years; and that you will watch 1 hour of TV per day throughout that time.

In short, my sums indicate that sticking with your old TV is a good idea unless you're happy to switch to something smaller. There are two clear winning options, each with a similar viewing experience and costing about the same over the 10-year period: a new energy-efficient 15-inch flat screen or a second-hand 14-inch CRT. Although the 15-inch flat screen has the lowest energy use, the 14-inch CRT wins

| | kg CO$_2$e | 0 | 200 | 400 | 600 | 800 | 1,000 | 1,200 |

Buy a 350 W 42-inch plasma screen for $750

Buy a 40 W 15-inch LCD for $300

Buy a 110 W 32-inch LCD for $450

Pick up a 14-inch CRT screen on Freecycle

Keep your old 28-inch CRT

Emissions from electricity use

Embodied emissions

FIGURE 3.1: The carbon footprint of different TV options, based on watching for 1 hour per day and not replacing again for 10 years.

overall at just 39 g per hour including the satellite receiver. But if you keep your TV for longer than 10 years, the winning option on every count is to buy the 15-inch LCD.

If you don't want to switch to a small screen, however, sticking with the 28-inch CRT screen is the best option, because the embodied energy of its manufacture has already been written off.

So the message is that although getting a new TV does give most people a chance to improve their energy efficiency, if you don't buy carefully, it is likely to do the reverse.

What about standby?

TVs typically use about 3 watts in standby mode, but since that probably accounts for at least 20 hours out of every 24, it means that your TV could well cause 15 kg (33 lbs.) of emissions over the course of a year even when there's nothing showing on the screen. If you have a small, efficient TV, that could be the biggest part of its annual footprint. Only you can decide whether standby adds enough quality to your life to justify the 15 kg (33 lbs.). My recommendation is that you cut it if you can, but don't let the issue torment you. If you spend a lot of time in front of the box, the additional exercise of switching off by hand will probably raise your quality of life slightly.

Lots of different devices around your home, all on standby at once, could collectively be more significant, and it should be said that some standby circuits use a lot more power than 3 watts. With a plug-in power meter costing about $15 you can check. No house should be without one.

Finally, it's worth mentioning that it also takes carbon to create the programs you watch—but that is a whole other story (see World Cup, page 158).

100 grams to 1 kilo (2.2 pounds)

A mug of tea or coffee

23 g CO$_2$e black tea or coffee, boiling only what you need
55 g CO$_2$e with milk, boiling only what you need
74 g CO$_2$e average, with milk, boiling double the water you need
236 g CO$_2$e a large cappuccino
343 g CO$_2$e a large latte

So if you drink four mugs of tea with milk per day, boiling just what you need, that's the same as a 60-mile drive per year in an average car. A single latte every day would be nearly 1 percent of the 10-ton lifestyle.

The shock here is the milk. If you take tea or coffee with milk, and you boil only the water you need, then the milk accounts for two-thirds of the total footprint (see Milk, page 72). The obvious way to slash the footprint of your tea is reduce the amount of milk, or simply to take it black (herbal tea, anyone?) (Figure 4.1). Although this will reduce your nutritional intake, you could easily replace the lost calories with something more carbon friendly such as a biscuit.

I have based my cappuccino and latte sums on the large kind that some of the coffee-house chains encourage you to quaff. These come in with a higher impact than four or five carefully made Americanos, filter coffees, or teas. They also mean you are drinking an extra cup of milk, perhaps without realizing it.

FIGURE 4.1: The footprint of one cup of tea or coffee with no sugar.

At my work we've suddenly decided that next week we're all going to do without milk in our drinks. At worst it will taste horrible. At best we'll change habits of a lifetime, resulting in decades of reduced hassle, lower carbon, slight cost savings, and possibly even fractionally improved health. It has to be worth trying.[*]

If you boil more water than you need (as most people do), you could easily add 20 g CO_2e to your drink. Boiling more than you need wastes time, money, and carbon; if you haven't yet developed perfect judgment, to avoid this you can simply measure the water into the kettle by using a mug.

Finally, think about your mugs. Buy sturdy ones; look after them and save hot water by only washing them up at the end of the day, rather than using a fresh mug for every cup.

---

[*] Update: we survived. It was horrible. I'm going to pick different battles. A little bit more herbal tea is drunk in the office these days, possibly as a result of the experiment.

A mile by bus

15 g CO$_2$e one of 20 passengers squeezed into a minibus
in the suburbs of La Paz
150 g CO$_2$e typical city bus passenger

The efficiency of a bus is just about proportional to the number of people it is carrying. It also depends on the amount of stopping and starting.

La Paz, Bolivia, is the place I think of where this principle is practiced to perfection provided you are prepared to set aside a bit of safety and comfort. Twelve-seater minibuses charge around town with 20 or more people crammed inside. You can get just about anywhere for one Boliviano—a few cents—and you are unlucky if you have to wait more than 5 minutes. Most people in the developed world would choose a luxury version of this for perhaps five times the price, but the principle is sound, and in Bolivia 10 years ago the "value proposition" met the market need perfectly.

All my numbers have factored in the fuel supply chains as well as the exhaust pipe emissions. I also include a component for the emissions entailed in manufacturing the vehicle, although for the bus this is a small consideration because they do so many miles before needing replacement.[1]

A diaper

89 g CO$_2$e reusable, line-dried, washed at 60°C (140°F)
in a large load, passed on to a second child
145 g CO$_2$e disposable
280 g CO$_2$e reusable, tumble-dried and washed at 90°C (190°F)
> So that's 550 kg (1,200 lbs.) per child for two and a half
years in disposables, the equivalent of nearly two and a
half thousand large cappuccinos.

Most parents will be relieved to hear that there is usually no carbon advantage to be had from reusable diapers. On average they come out slightly worse, at 570 kg (1,250 lbs.) per child compared with 550 kg (1,200 lbs.) for disposables. And if you wash them very hot and tumble-dry them, reusables can be the worst option of all. However, if you put your mind to it, you can make reusables the lowest-carbon option. To do this, pass them on from child to child (so that the emissions embedded in the cotton are spread out more), wash them at a lower temperature (60°C/140°F), hang them out to dry on the line, and wash them in large loads.

For a disposable diaper, most of the footprint comes from its production. But about 15 percent arises from the methane emitted as its contents rot down in landfill (contrary to the myth that if you wrap them up in a plastic bag they will never rot at all).

The study I'm basing my figures on assumed that the average child stays in diapers for about two and a half years and is changed just over four times a day.[2] On this basis, in the U.K., diapers account for something like one two-thousandth of total greenhouse gas emissions—or more like half a percent for homes with babies.

What does all this mean for the carbon-conscious family? If you have two children and stick to non-tumble-dried reusables throughout, you might be able to save nearly half a ton CO_2e. You will also cut out landfill. It's a significant efficiency, but (here's the catch) you need to know your own minds before you start out because if you give up, revert to disposables and trash the reusables, it could be the option with the highest footprint of all. But try to keep all of this in perspective: if you take just one family holiday by plane, you will undo the carbon savings of perfect diaper practice many times over.

When he was the U.K.'s climate change secretary, Ed Miliband recently drew on the same diaper study to defend his announcement that his own children wear disposables. He was roasted—somewhat unfairly I thought—by blogging eco-mums who claimed that the study was fatally flawed. Poor chap. At least he'd thought about it. The debate illustrates, yet again, that this kind of analysis is more murky and subjective than we might think.

A basket of strawberries

150 g CO$_2$e (or 600 g per kilo/270 g per pound) grown in season in your own country

1.8 kg CO$_2$e (or 7.2 kg per kilo/3.3 kg per pound) grown out of season and flown in, or grown locally in a hothouse

> How have we got into the habit of buying tasteless out-of-season strawberries, which have a footprint more than 10 times the tastier seasonal version?

Although I've given just one number for local, seasonal straw-berries, the precise footprint depends on such things as the soil, the use of fertilizer and the use of polytunnels.[3] Some of these variables increase both the yield and the emissions per acre, so whether they result in more or less carbon per strawberry is not so simple to work out. Luckily, they are all so much better than the out-of-season version that a good enough rule of thumb is just to stick to those grown in your own country—unless your government subsidizes the heating of greenhouses (as is the case, for example, in the Nether-lands). This kind of hot-housing is, broadly speaking, just as bad as air-freighting the fruit from hotter countries (see Fly-ing, page 133, and Asparagus, page 85).

In short, then, the best advice is to wait until they are in season, then enjoy them twice as much. Or if you really can't wait, buy frozen or tinned: these lie somewhere in the mid-dle of the range, in carbon terms, along with those traveling "middle distances" by road and boat from warmer climes.

All the figures here have taken account of the 23 percent average wastage between the field and the checkout. A small amount of the footprint is the packaging and this is actually in a good cause if it enables more of the strawberries to find their way into our mouths. The footprint of the plastic will typically be lower than that of the wasted fruit.

A mile by train

150 g CO_2e Intercity standard class
160 g CO_2e London Underground
190 g CO_2e light rail or tram
300 g CO_2e Intercity first class

> An 18-mile intercity rail journey has the same footprint as a cheeseburger, whereas a mile and a half journey on light rail is equivalent to a cup of milk

Although trains can be a relatively green way to get around, the figures above show that the emissions of rail journeys are higher than you might think. All the numbers provided include the direct emissions and electricity consumption of the moving train itself but also attempt to take account of the embodied emissions from train manufacture, the upkeep of the rail network and the running of all the infrastructure.[4]

The amount of energy required to propel a train down a track depends mainly on just a few simple things:[5]

> How fast the train goes. The air resistance goes up with the square of the speed.

> How many stops there are. Each stop wastes energy—the exact amount being proportional to the square of the speed and the weight of the train. Some newer trains reduce this stoppage waste through "regenerative braking," a similar technology to the one used in hybrid cars.

> Rolling resistance of the wheels on the track. This is lower for trains than for cars because metal wheels on metal tracks are more efficient than rubber tires on asphalt. The rolling resistance goes up proportionally with the weight of the train.

> The type of fuel used. Electricity beats diesel because although there are inefficiencies in generating it from fossil fuels in the first place, once this has been done the train engine can turn almost all of the power into movement. A diesel engine is much less efficient.

Long-distance Intercity trains go fast (that's bad) but stop infrequently (that's good). In the U.K., they're often electric (that's good), but they're also extremely heavy (that's bad). The weight of the train per passenger seat, amazingly, is around twice that of an average car. Just to be clear, what I am saying is that the weight of a full train is twice that of all the cars that would be needed if each passenger drove instead. Professor Roger Kemp,[6] who has looked at this astonishing fact in detail, explains it in terms of over-engineered safety: trains weigh at least twice what they need to because we have become obsessed with safety and have forgotten that rail travel is already over 100 times safer than driving. A couple of miles from my house an Intercity train derailed and rolled down a high embankment. Incredibly, only one person was killed. The event was still splashed across the national news, raising public fears, even though so many more people die on the roads every single day. One price of this excessive focus on safety may well be that twice as much energy is required to get our trains moving every time they leave a station.

First-class travel deserves a mention because the number of seats you can squeeze into a first-class carriage is around half the number in a standard-class carriage. This means that the weight being moved per person is doubled again; we're now up to the weight of four cars per seat. I sometimes board trains where half the length is nearly empty first class and the rest is crowded standard class, suggesting that the real weight being hauled per first-class passenger may be even higher.

Things are a bit more complex when it comes to the Eurostar, because when it's in France it runs on electricity that comes predominantly from nuclear power. This is low-carbon energy, whether or not you think nuclear power is worth it in other ways. However, I don't think it is useful to think of trains in nuclear-friendly France as having a smaller footprint than those elsewhere—which is how they are sometimes portrayed. That's because all the nuclear electricity that French power stations can produce would get used up regardless of whether any trains were running. In that sense, the trains are effectively powered by the fossil fuel plants that provide the extra electricity over and above the nuclear "baseload" (see A unit of electricity, page 56, for more on this somewhat confusing concept of marginal depend).

Interestingly, the London Underground is almost as low-carbon, per passenger mile, as Intercity trains, despite stopping much more often. This is mainly because people are packed in so tightly—almost tessellating, nose to armpit. Other reasons are that the Tube travels relatively slowly, is all-electric, and has lighter trains.

Overall, trains are generally a lot greener than cars but not as good as walking, cycling, or staying at home. A sensibly designed car can win, provided you fill it with people. Even two people traveling together are better off driving an efficient car than traveling first class. (See also New York City to Niagara Falls return, page 116.)

A 500 mL (16 oz.) bottle of water

110 g CO₂e locally sourced and using local distribution
160 g CO₂e average
215 g CO₂e traveling 600 miles by road
> A bottle a day would add up to 0.6 percent of the 10-ton lifestyle.

Bottled water is more than 1,000 times more carbon intensive than its tap alternative, so knocking it out of our lives has got to be a simple win. It doesn't even taste better.

Processing the water is the easy part: the bulk of the emissions come from packaging and transport. There is 80 g CO_2e per quart just for the plastic. On top of that is the energy required to melt the PET (polyethylene terephthalate) balls down and mold them into bottles. Transport is significant because water is so heavy. If it has gone 600 miles by road, that could add a further 115 g CO_2e per bottle.[7] Shipping from Europe to North America is clearly not good news.

The small town of Bundanoon in Australia is world leader in the fight against the stuff with a ban of bottled water already in place. Concord in Massachusetts has announced plans to go bottled water free from January 2011, despite threats from the bottled water industry to sue. Meanwhile London has announced plans to start reintroducing public drinking fountains. All these are encouraging steps forward. If everyone switches away from bottles, it will be great for the environment and still just as healthy, refreshing, and convenient. Interestingly, even though people will be financially better off, the economy may look as though it has slowed down a fraction. This is a nice illustration of how inadequate it is to measure how we are doing by our economic growth. When we are all using the fountains, we might collectively look a shade poorer on paper because the few people who make their living persuading us to buy the bottled stuff will need new jobs. But that will be more than compensated for by the extra cash that the average person will save. So the economy will recede as we all get better off. Let's not cry for the peddlers of bottled water either. Even if you don't believe that they had it coming to them, they are clearly talented and persuasive people who are also more than capable of being successful in constructive careers.

If the world consumes 53 billion gallons of this bottled water per year,[8] that's 80 million tons of greenhouse gases, or one-sixth of a percent of global emissions. This is a win worth having!

A letter

140 g CO_2e a 10 g letter made from recycled paper and recycled by you
200 g CO_2e a typical 25 g letter printed on virgin paper and sent to landfill
1,600 g (3.5 lbs.) CO_2e a small catalogue sent to landfill
> If you have five letters delivered per day plus two catalogues per week, that's a massive 480 kg (1,060 lbs.) CO_2e per year, nearly 5 percent of the 10-ton lifestyle.

Mail clocks up a carbon footprint in four basic ways (Figure 4.2):

> Paper production. The carbon footprint of paper manufacture depends on the recycled content, the quality of the paper and the efficiency of the mill. The junk mail coming through our door generally uses high-quality stuff and doesn't tend to boast any recycled credentials. My estimates are based on paper that has a less than one-fifth recycled content. That gives it a footprint of 2.35 kg CO_2e per kilo (1.07 kg CO_2e per pound). The best estimate for pure virgin paper comes in at 2.59 kg per kilo (1.18 kg CO_2e per pound), and 100 percent recycled paper at about half of that; it takes about half as much energy to create new paper from old paper as it does to create paper from trees.[9]

> Printing. For the footprint of printing on the paper to turn it into glossy and enticing sales literature, I estimate an additional 350 g CO_2e per kilo (160 g per pound).

> Postage. For a standard letter, this accounts for most of the footprint. It's impossibly difficult to trace the carbon footprint of mailing a letter by direct means. However, if you take the footprint of the postal services sector as a whole and divide it by the turnover of that sector, you can get a broad idea of the carbon footprint per unit of cost. In the U.K. it comes out about 380 g CO_2e per £1 spent (250 g per US$1). A 25 g regular mail letter would have cost 32p (48 cents) in the U.K., and we can associate a carbon footprint of about 120 g CO_2e with that. So most of the impact of a junk letter comes from the burden that it places on the whole infrastructure of the postal system: vans, trains, and sorting offices.

> Decomposition. A good deal of junk mail ends up in landfill, where it decomposes anaerobically and produces methane. For this I have allowed 550 g CO_2e per kilo (250 g per pound) of paper.[10] You can prevent this, of course, by recycling as much mail as possible. This is OK to do even if the letter has a plastic window. But do remove any other plastic—such as film wrap.

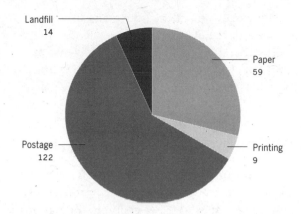

FIGURE 4.2: The carbon footprint of a 25 g letter, printed on virgin paper, sent by mail, and thrown into landfill (grams CO_2e).

Eliminating junk mail will declutter your life as well as save carbon. The purpose of most of it is to persuade you to buy stuff you don't need, so brain purification is probably the biggest reason of all for putting an end to it. Shelly Shumacher writes in ehow.com[11] that Americans receive an average of 41 pounds of junk mail per year and 44 percent of this ends up in landfill, and she also offers advice on how to get the stuff out of your life.[12]

Finally, a message to the instigators of junk mail: more and more people will think badly of you for using high-carbon marketing techniques. If you must use mailouts, at least keep your databases clean, use recycled paper, and keep your messages short.

Sending an email beats sending a letter hands down (see page 20).

1 kg (2.2 lbs.) of carrots
0.25 kg CO_2e local, in season
0.3 kg CO_2e average
1 kg CO_2e shipped baby carrots
> So a bag of carrots is like a 2-mile train ride.

At around 2 g CO_2e per calorie, these and other root vegetables are some of the most climate-friendly foods available—and healthy too. If you ate only these foods and others that have similar carbon intensity, you could feed yourself for just over 1 kg CO_2e per day, or less than 500 kg CO_2e per year.

Seasonal vegetables have small carbon footprints because they avoid all of the main greenhouse gas sources for food: they are grown in natural conditions without artificial heat, they don't go on airplanes, and they don't incur the inefficiencies inherent in the production of food from animals.

If you go on to boil your carrots for 10 minutes, you will add a few more grams CO_2e per pound to the footprint. (For more on cooking, see Boiled potatoes, page 71.) My children will only eat their carrots raw. That suits me fine. It's better from every angle—there's less carbon emission, it saves time, and the nutritional value is better.

Note that some baby varieties have a much lower yield per acre of land, resulting in higher emissions per pound. So it usually makes sense to buy full-sized, classic varieties. And, as with other vegetables, favoring misshapen specimens may help prevent wastage in the supply chain (see page 182).

A newspaper

0.3 kg (0.66 lbs.) CO_2e the U.K.'s *Guardian Weekly,* recycled
0.40 kg (0.77 lbs.) CO_2e *Globe & Mail* Saturday edition, recycled
0.41 kg (0.91 lbs.) CO_2e *New York Times* weekday edition, recycled
0.43 kg (0.94 lbs.) CO_2e *Globe & Mail* weekday edition, recycled
0.8 kg (1.8 lbs.)CO_2e the U.K.'s *Guardian Daily,* recycled
1.5 kg (3.3 lbs.) CO_2e *New York Times* Sunday edition, recycled
3.2 kg (7.1 lbs.) co2e *New York Times* Sunday edition, chucked into landfill
4.1 kg (9 lbs.) CO_2e a typical British weekend broadsheet paper, sent to landfill
> I estimate that the *New York Times* every day, including Sundays, adds up to 207 kg (455 lbs.) CO_2e per year, if you recycle them all, or

447 kg (984 lbs.) CO_2e, if you chuck them in a garbage bin and off to landfill. The latter is equivalent to flying from New York to Atlanta and back or from San Francisco to Vancouver.

It's amazing how energy-hungry newspaper production can be. And the figures provided here are on the low side, because none of them take account of the footprint of journalism itself—including the newspaper offices and staff flights. At the highest end of the spectrum, just the *New York Times* on Sunday each week could add up to almost 2 percent of a 10-ton lifestyle if you don't recycle it (Figure 4.3). At 1.25 kg (2.8 lbs.), the weight of a U.K. weekend broadsheet is part of the problem. In our house only a tiny fraction would be read, so the rest might as well never have been printed.

The reasons why recycling is so important are twofold. First, if paper is disposed of in landfill sites, it emits methane as it rots. Second, for each newspaper that isn't recycled, one

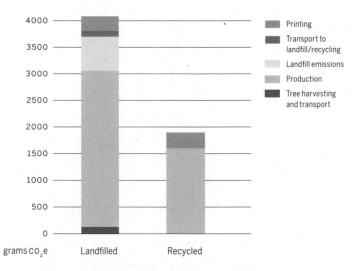

FIGURE 4.3: The carbon footprint of a weekend newspaper. Sending paper to landfills causes methane emissions *and* means that more carbon-intensive virgin paper has to be produced.

more newspaper's worth of virgin paper has to be manufactured. For these reasons, throwing your paper in the general waste more than doubles its footprint.[13]

Opting for a slimmed-down weekly paper is one good way to reduce both emissions and clutter. We take the U.K.'s *Guardian Weekly*, weighing just 300 g—condensed and interesting, if a few days out of date. Another way is to get your news online. If you do this for an hour a week on a 50-watt laptop and if we multiply that by, say, five to take account of the production of the laptop, the running of your network, and the electricity consumed by all the hubs and servers around the world that support the websites you browse, it still comes to around half the impact of the *Guardian Weekly*. If only I could take my laptop into the bath...

A pint of beer

300 g CO_2e locally brewed cask ale at the pub
500 g CO_2e local bottled beer from the store or a pint of imported beer in a pub
900 g CO_2e bottled beer from the store, extensively transported
> A pint of local ale per day in the pub would be 1 percent of the 10-ton lifestyle. A few bottles of imported lager per day might be as much as 10 percent.

The good news is that North America's robust microbrewery industry gives many people plenty of tasty lower-carbon options.

The beer at the low end of the scale is based on figures for the Keswick Brewing Company, a microbrewery quite near where I live. Just about everything you can think of was included in the study I did for them (Figure 4.4). There were the obvious things such as ingredients, packaging, fuel, electricity, and transport. I also included such elements as staff travel, the carbon cost of having to replace their equipment every so many years, and office stationery.

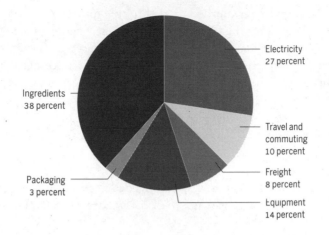

FIGURE 4.4: The footprint of cask beer from the Keswick Brewing Company.

For the Keswick Brewing Company, I estimated that ingredients accounted for about one-third of the footprint, fuel and electricity about another one-quarter, and staff travel about one-tenth. The fermentation process itself releases CO_2, accounting for about 15 g per pint, but these don't count, since that carbon was absorbed by the ingredients as they grew. Most of the company's beer is sold in reusable casks, so the footprint of packaging is kept right down. Distribution is 7 percent, even though all their deliveries are fairly local, because beer is such heavy stuff.

A few miles from the Keswick Brewery is another, larger brewery. Delivery from there to pubs just down the road is via a distribution center in Wolverhampton, a couple of hundred miles away. This is the usual story for big breweries and their subsidiaries. Even the country of origin is not always obvious from the branding. Although a few hundred road miles are not usually the most significant factor for foods, beer is an exception because it's so heavy. Hence opting for local ale is usually a good idea.

For home consumption, and thinking for a moment only of carbon rather than taste, cans are slightly better than bottles, provided you recycle them. (I can feel the connoisseurs at Keswick cringing as I write.) Heeding this advice is especially important if the beer is traveling a long way because the glass also adds to the weight.

Wherever and whatever you drink, a single pint of a quality beer is almost always better for both you and the world than spending the same money on several tins of bargain-basement brew.

Finally, as though it were needed, the carbon angle gives yet another reason not to drink and drive (see Car crash, page 139).

A bowl of porridge

82 g CO₂e traditional Scottish, made with oats and water only
300 g CO₂e about half milk (just how I like it)
550 g CO₂e milky and sweet
> So a bowl of traditional Scottish porridge is equivalent to a 90-second cell-to-cell phone call.

A bowl of half-milk porridge every day would be about 1 percent of the 10-ton lifestyle. Cement and porridge made like this don't just look the same; they also have very similar carbon intensity per kilo.

Oats, like most cereals, is a fantastic low-carbon food that also happens to be healthy and tasty. If you fed yourself entirely on milky sweet porridge, it would cause just 900 kg CO_2e per year. By sticking to the Scottish water-based version, you'd cause a trifling 340 kg CO_2e per year—about one-tenth of the typical U.K. diet. As with a cup of tea (see page 37), it's the milk that dominates.

The cooking is about half of the footprint of the traditional Scottish version. I've assumed that you cook it on the stove and never have the lid on because you are stirring like crazy, trying to save yourself a nightmare of cleanup. A nonstick pan should help. Better still, the microwave is lower carbon than an electric burner or gas ring and doesn't cause sticking; but keep a close watch or it will turn into an exploding mess from Doctor Who. Enough said about all this. I am the last person who should be writing a cookery book.

A bowl of cold breakfast cereal or granola prevents the cleanup nightmare and offers similarly excellent carbon per calorie to a bowl of porridge with just a bit of milk.

A shower

90 g CO₂e 3 minutes, efficient gas furnace, aerated showerhead
550 g CO₂e 6 minutes in a typical electrically powered shower
1.9 kg (4 lbs.) CO₂e 15 minutes in an 11-kilowatt, high-volume, electrically powered shower

> If you have high-carbon shower habits, there could be half a ton per year to be saved here—equivalent to a return flight from San Francisco to Vancouver, Chicago to Dallas, or New York to Atlanta.

At the low end of the spectrum, 3 minutes is how long I take if I wake up half an hour before my train is due to leave. Gas is a more efficient way of providing heat than electricity, provided you have a reasonably efficient furnace. The aerated shower head helps by making less water feel like more. In theory at least, it saves water and carbon without your having to forgo any comfort at all.

If you are in a family of four and you each spend 15 minutes in an electric shower every day, you may be able to reduce your household footprint by a ton per year just by switching to an aerated showerhead. Switch to a gas-powered shower, and there's another half ton to be saved. Finally, you can cut the remaining emissions by a factor of three by having 5-minute showers—and it is only at this point that you are having any impact on your lifestyle. You will be swapping time in the shower for time doing almost anything else that you want: reading a book, lying in bed, both of these at once, or whatever you like. If you take all these measures, your family could knock off 2 tons per year—and save about $500, which would easily pay for the couple of paperbacks you might each luxuriously read in bed during the time you have liberated.[14]

The showers in Iceland are worth a mention as the most luxurious I've ever had. Geothermally heated and almost zero CO_2e, they are all the more enjoyable after a day out in the abundant rain and snow there. Unfortunately you have to fly to get there. (See also Bath, page 83.)

An ice cream

50 g CO_2e a 60 g Popsicle from the supermarket, eaten on the day of purchase
500 g CO_2e a big dairy ice cream from a van

The Popsicle is essentially frozen sugary water, and in the supermarket the refrigeration is likely to be relatively efficient.

At the high-carbon end the dairy ice cream's footprint is higher for three reasons: it's bigger, it's dairy based, and it's been kept·cold in a much less efficient mobile refrigeration unit. The inclusion of dairy ingredients means that all the inefficiency of ruminant livestock farming has been incurred. My figures are just based on cigarette-packet calculations. I've guesstimated from a broad understanding of the footprint of different food ingredients and transport impacts and from knowing a little bit about mobile refrigeration.

A unit of heat

50 g CO_2e using a solar water-heating panel
244 g CO_2e using a modern (90 percent efficient) gas furnace
400 g CO_2e using an old, 55 percent efficient gas furnace
660 g (1.45 lbs.) CO_2e from U.S. grid electricity
1,060 g (2.34 lbs.) CO_2e from Australian grid electricity

> By a "unit" I mean 1 kilowatt-hour. That is enough to run a
  "one-bar" electric fire for 1 hour or enough to boil about 4 gallons
  of water in an electric kettle.

At the low end, the solar water-heating panel has no opera-
tional emissions. I haven't given it a zero CO_2e rating because
the manufacture of the panel itself will have a carbon footprint. The exact number depends on factors such as the design
of the device, where it is used, and how it is maintained, so
my figure is really just a guesstimate. One problem with solar
heating is that it tends to be "low-grade" heat. In other words
it's all right for warming up baths and gently heating rooms
but not usually much good for boiling kettles or making toast.

In the middle of my scale is heat generated by an efficient
gas-fired furnace, such as might power a new central heating
system. In this scenario your heating is done by fossil fuels,
but at least you're using them fairly efficiently: the only losses
will typically be around 10 percent, as the energy disappears
out of the flue.[15]

(That said, in the case of a central heating system there
may also be inefficiencies caused by heating rooms that you
are not actually using. If the only room you want to heat up
is the kitchen, the most efficient thing you can do may be to
turn on the gas stove. That way nothing goes up the flue, and
all the heat goes into the room you want to keep warm.)

At the high end of the scale is electricity. This is a "high-
grade" form of energy that can be used for many different things, so it's generally a waste to use it just for heating.
The precise footprint depends on which country you are
in (see page 162), but with only a few exceptions the figure
will always be high because the electricity is usually generated from fossil fuels and—unlike with a gas furnace in your
home—more than half the energy in the fuels is lost in the
power station or transmission grid. In other words, it's generally inefficient to use electricity just for heating. In the U.K.,
the average unit has a footprint of about 600 g (1.3 lbs.) CO_2e,[16]

whereas the figure is higher, for example, in coal-dependent Australia.

A unit of electricity

60 g CO_2e from the Icelandic grid
220 g CO_2e from the Canadian grid
600 g (1.3 lbs.) CO_2e from the U.K. grid
660 g (1.45 lbs.) CO_2e from the U.S. grid
900 g (1.98 lbs.) CO_2e from the Chinese grid
1,060 g (2.34 lbs.) CO_2e from the Australian grid

> The carbon impact of using an additional unit of electricity is often higher than we're led to believe.

Electricity generation is one of the principal causes of carbon emissions all over the world. However, as we've seen, the exact carbon cost of each unit of power depends on the precise mix of generating fuels used in your country. Icelandic electricity comes almost exclusively from fossil fuel-free geothermal and hydropower plants, so the only footprint comes from creating and maintaining the infrastructure.

Australian and Chinese electricity, by contrast, comes mainly from highly polluting coal. U.S. and U.K. electricity is somewhere in the middle, coming from a mixture of coal, gas (which is less polluting than coal but is still a fossil fuel), nuclear (which has a low carbon footprint but is contentious in other ways), and a smattering of renewables. The mix is significantly cleaner in Canada but varies hugely between provinces.[17]

Most people who think about carbon footprints are used to the idea that each unit we consume causes a fixed quantity of CO_2 emissions. However, the truth is somewhat more complex than that. A more meaningful way to think about the carbon footprint of your own electricity use is to think of it as being additional to all the power consumption that was already going on before you flicked on the light or appliance. Looked at this way, the extra demand that you place on the

Country	Direct emissions from power generation, per unit consumed kgCO$_2$e per kWh	Estimate of total footprint per unit consumed* kgCO$_2$e per kWh	Estimated footprint of marginal demand** kgCO$_2$e per kWh
Australia	1.00	1.06	1.06
China	0.84	0.90	1.06
Iceland	0.00	0.06	0.06
Norway	0.00	0.06	1.06
U.K.	0.54	0.06	1.06
U.S.	9.60	0.66	1.06

* (including the carbon cost of extracting fuel from the ground, maintaining and building power stations, wind turbines, etc.)

** e.g. the carbon cost or savings of each unit of electricity you choose to use or save

TABLE 4.1: The carbon footprint of electricity consumption in different countries. The marginal demand column shows that, unless you live in Iceland, someone somewhere is likely to have to burn more coal if you use more electricity.

grid is met entirely through additional fossil fuels, because the renewables in your country will already be running at full capacity. In other words, when you turn the lights on, you don't personally affect the amount generated by renewables because they are already going flat out. Rather, what you trigger is almost certain to be a lump of coal thrown into a power station. This is the case throughout Europe, because even in countries where all electricity comes from renewables or nuclear, adding to demand reduces the amount of electricity that those countries are able to export, thereby increasing the fossil fuel generation in other nations. In terms of "marginal demand" (see Table 4.1)—each unit of electricity you consume has a footprint of at least 1 kg (2.2 lbs.) CO$_2$e per unit, regardless of which country you live in.

One exception is Iceland, where, for the moment at least, it does look as though you can more or less use as much electricity as you like without boosting your footprint. The country is overflowing with hydroelectric and geothermal power: you can *see* the energy almost everywhere you go, boiling out of the mud and pouring over waterfalls. But once Iceland works out how to export its clean energy, or how to import enough of the world's heavy industry to use up the renewables capacity, electricity will become a scarce resource for Icelanders, just as it is for the rest of us. In the meantime, enjoy it!

The great green tariff swindle?

"Green" or "renewable" electricity tariffs and suppliers may sound attractive, but the hard reality is that signing up is unlikely to reduce the climate change impact of your electricity significantly. This applies whatever the color of the company's logo or however ecological the company name might sound.[18]

The two main claims made by the "green" providers are that electricity comes from renewable sources and/or that they use the money you spend on your bills to invest in a new renewable capacity. Neither of these is necessarily what it might appear.

The "from renewable sources" claim

All electricity suppliers in the U.K. are obliged to submit Renewable Obligation Certificates (ROCs) to the government for up to 10.4 percent of the electricity they sell to their customers. They can get these certificates either from generating their own renewable power or by buying them from others. Suppose a company has a tariff in which all electricity is sourced from renewables. It sounds great. However, this means that the supplier gets a lot more ROCs than they need to hand over to the government. The normal practice is for the "green supplier" to sell these to other suppliers, thereby allowing them to simply source less of their own power from renewables. So the net carbon benefit is zero—but the "green supplier" stands to benefit because it has managed to charge you a premium. The tariff only makes a difference to the extent that the provider retires some ROCs (tears them up) instead of selling them on. In the U.K., energy supplier Good Energy claims to do this with 5 percent of them, although this has been challenged, with some people suggesting they have been retiring only 2 or 3 percent. It doesn't much matter, because they are arguing over such low percentages. The main point is that well over 90 percent of the ROCs are kept in circulation. If you switch to the "green tariff" offered by one of the larger electricity suppliers, the chances are that no ROCs are retired at all, and you are allowing them to worsen the energy mix in their other tariffs, while using the "green" story line as a way of charging you more.

The "investing in renewables" claim

What if a company claims that it will invest so much of every dollar you spend via the company in new wind farms and other renewable energy projects? This sounds great, but what it could boil down to is that the supplier is simply engaging in two different business activities. One is being an electricity provider just like all the others, but with "green" branding. The other part of the enterprise is investing in renewable power generation. Both of these could potentially be good business opportunities regardless of any environmental considerations. The key question is whether the investments in the new wind farms would still be made if you got your electricity from elsewhere. Is the company promising to invest to the tune of your electricity expense in projects that would otherwise not go ahead at all? In other words, is it genuinely *additional*? This is a very long way from being clear to me. One thing that is certain is that it is possible to run a roaring commercial enterprise along these lines.

I am not saying that the companies claiming to provide greener electricity aren't greener than average. They probably are, and I do get my own electricity from one of them. What I am saying is that their impact may not be quite as low as you think. The overall message is that if you want to reduce the footprint of the electricity you buy via the grid, the only real way to do it is to consume less.

Spending $1

Minus 220 kg (485 lbs.) CO_2e on a well-executed rainforest preservation project
Minus 2 kg (4.4 lbs.) CO_2e on solar panels
105 g CO_2e on financial, legal, or professional advice
480 g CO_2e on a car
620 g (1.37 lbs.) CO_2e on a typical supermarket cart of food[19]
3.1 kg (6.83 lbs.) CO_2e on flights[20]
6.5 kg (14.3 lbs.) CO_2e on gasoline
4 kg (8.8 lbs.) CO_2e on the electricity bill
10 kg (22 lbs.) CO_2e and beyond on budget flights

Unless you are deliberately investing in something that reduces emissions elsewhere, it is just about impossible to spend money without increasing your carbon footprint. Everything causes ripples of economic activity and, with it, emissions. So with wealth comes carbon responsibility. I'm hardly the first person to have suggested this, but it's an important concept. So what are you going to do with your $1?

If all your money goes into travel, you may be at the worst end of the irresponsibly wealthy. If you invest it in forests and wind farms, you are at the opposite extreme, using your wealth to bring about a low-carbon world. If you spend a million dollars on fine art, you are mainly passing on the responsibility for doing the right thing with that cash to the artist or the dealer. If you stick it under the mattress, it is doing neither harm nor good.

Of the specific examples given above, flying gets to be such a high-impact way of spending cash for two reasons. First, the aviation industry can buy its fuel for around 50 cents per liter ($1.89 per gallon). Second, it then burns it at an altitude where it has, as a best estimate, nearly twice the climate change impact that it would have had at ground level. Leaving the lights on is another of the cheapest ways of trashing the planet, suggesting that for all the talk of higher fuel prices, we are a long way from establishing a serious financial incentive to go green.

My gasoline figure is based on $2.70 per gallon. My sums take account of the extraction, shipping, and refining of the fuel but not the depreciation or maintenance of the car.[21]

At the positive end, I have included some of the fairly limited options for actually doing carbon-friendly things with money. They range enormously in their effectiveness, which is something not all policy makers seem to have fully grasped.

In our input–output model, we've looked at the carbon intensity of industries from agriculture to manufacturing to education and social services. No industry sector in our model comes in below 100 g per U.S. dollar. The more you think about this, the clearer it becomes that there is simply no avoiding the advantages of slowing the economy down or of changing its structure. We could do with spending less time charging around earning as much as we can to buy things we don't really need. We would do well to become better at enjoying what we've got—and to disentangle our self-esteems from our wages. Without wishing to sound like a sandal-wearer, I think it's clear that we've become locked into a mindset that is not going to serve us well over the coming decades. If you're not convinced, have a read of Tim Jackson's book *Prosperity without Growth*.[22]

1 kg (2.2 lbs.) of trash

200 g CO_2e garden waste
700 g (1.54 lbs.) CO_2e average trash contents
9 kg (19.8 lbs.) CO_2e aluminum and copper

The average U.S. citizen sends 570 kg (1,250 lbs.) to landfill or incineration[23] each year and recycles just 290 kg (640 lbs.).

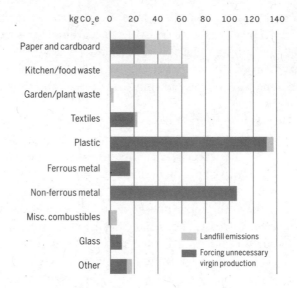

kg CO$_2$e

FIGURE 4.5: The annual footprint of the average U.K. homeowner sending waste to landfill rather than to recycling or composting.[25]

This causes around 400 kg (880 lbs.) CO$_2$e, or 4 percent of a 10-ton lifestyle.[24]

By "trash" I mean things you dispose of by putting them in the normal garbage as opposed to recycling or composting them. Looked at this way, the footprint has two parts.

First, there are the landfill emissions, which are due mainly to stuff rotting down underground, without air. This anaerobic decomposition produces methane, only some of which gets captured, and the rest escapes to warm the world. (This isn't an issue for metals, glass, and building materials, of course, because they don't rot down in the way that food, paper, and garden waste do.) There is also a little bit of fossil fuel required to run a landfill site.

Second, there is the fact that by not recycling something, you are forcing more virgin materials to be produced for use in future products. This isn't an issue for food, for which

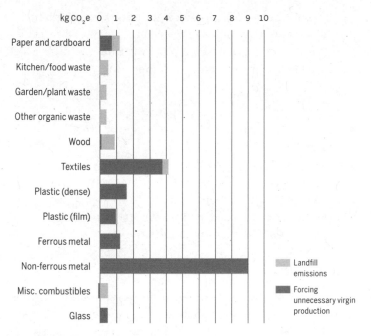

FIGURE 4.6: The footprint per kilo (2.2 pounds) of throwing stuff into landfill compared with recycling or composting it. In other words, this graph shows the *difference* that recycling makes.

recycling was never an option. But for metals, textiles, plastics, and paper it is a big deal.

Figure 4.5 shows that recycling our aluminum and plastic is where most of us can make the biggest improvements. That's mainly because it takes so much more energy to make a brand-new aluminum can or plastic bottle than it does to make a new one from an old one. Kitchen waste is a key area, too, because of the large amount of methane it produces when it rots underground.

Figure 4.6 shows that when you are standing with 1 kg (2.2 lbs.) of something in your hand, if that something is aluminum, it is particularly important that you recycle it. The next most important per kilo are textiles.

The significance of food waste is underplayed in both these graphs because they don't take account of the footprint of needless production. These graphs just show the difference between landfill and recycling or composting.

Doing the dishes

Almost zero CO_2e (but the plates aren't clean) by hand in cold water

540 g (1.2 lbs.) CO_2e by hand, using water sparingly and not too hot

770 g (1.7 lbs.) CO_2e in a dishwasher at 55°C (130°F)

990 g (2.2 lbs.) CO_2e in a dishwasher at 65°C (149°F)

8 kg (17.6 lbs.) CO_2e by hand, with extravagant use of water

> Running a dishwasher twice a week on the economy setting comes to 80 kg (176 lbs.) per year, equivalent to a 110-mile drive in an average car.

The results of the great dishwasher versus handwashing debate are as follows. The most careful hot-water handwashing just about beats the dishwasher but loses out badly on hygiene (nearly 400 times the bacteria count on the dishes) and time (four times as long as loading the dishwasher). Overall the dishwasher wins, particularly because the figures here probably don't reflect the most energy-efficient machines that are now on the market. I also haven't included the carbon savings that are possible if you set your machine to run in the middle of the night when electricity demand is low and the grid becomes more efficient.

The handwashing figures are based on a study of people around Europe,[26] but I've used the U.K. electricity mix to calculate the carbon. (If you live in nuclear powered France, don't be fooled into thinking your electricity consumption doesn't matter so much. It all gets traded around, as discussed on page 56).

My figures for the dishwasher are based on always running a full load, and they include 130 g CO_2e for the wear and tear on the machine itself (based on a fairly expensive "built to last" model that you keep for 10 years[27]). The conclusion, then, is get a dishwasher. It simultaneously helps the planet, your health, and your lifestyle. When you buy one, choose a make that will last, and look after it. Try to always run it full, use the economy setting when possible, and run it in the middle of the night if you can, because the electricity is less carbon intensive.

I haven't included anything for the detergent or the domestic water consumption, because they are nothing compared with the impact of heating the water.[28]

A final note: I have known people routinely wash their stuff by hand before putting it in the dishwasher. This must be the worst of all options and ranks alongside ironing your spouse's socks for needless slavery (see Ironing, page 25). If this is your routine, please consider yourself liberated.

A roll of toilet paper

450 g CO$_2$e recycled paper
730 g (1.6 lbs.) CO$_2$e virgin paper
> If you have typical North American wiping habits, that comes out at 75 kg (165 lbs.) CO$_2$e per year or three-quarters of a percent of the 10-ton lifestyle.

The typical North American supposedly uses 57 sheets of toilet paper per day. That seems excessive to me, although I haven't been counting. The figure comes from the ToiletPaperWorld.com website, and surely they must know these things.[29] The Worldwatch Institute puts annual consumption at 23 kg (50 lbs.) per year for a North American, 1.8 kg (4 lbs.) per year for an Asian, and just 400 g for the average African.[30]

I'm not sure I want to launch into a detailed exploration of bathroom technique here, but because three-quarters of a percent of the 10-ton life seems high for such a simple and brief part of our lives, it does seem worthy of a moment's personal reflection. My numbers show that a sense of economy is in order. If, as I suspect, many of us could halve our usage without any negative side-effects, then it's an easy and worthwhile carbon win.

I'm not advocating hardship here— just calling for a simple perspective check; are our backsides in their rightful place, or are they getting spoiled? Have decades of ads talked us into believing that a pampered bum is one of the hallmarks of a rich and fulfilled life? My footprint figures here are based on numbers from Tesco, whose research suggests a carbon cost of 1.1 g per sheet for their recycled stuff and 1.8 g for traditional paper. So that's three spam emails for a sheet of recycled,[31] five for virgin, or two sheets of virgin for one genuine email.

Driving 1 mile

350 g CO_2e a Fiat 500 doing a steady 60 miles per hour
850 g (1.9 lbs.) CO_2e an average U.S. car achieving a typical 22.4 miles per gallon
2,500 g (5.5 lbs.) CO_2e a Mercedes Benz SUV, new but not looked after, doing 90 miles per hour
> So driving a vehicle 10,000 miles would use between 35 and 250 percent of the 10-ton lifestyle, depending on what you drive and how you drive it.

At the low end of the scale, for four people traveling together in a well-maintained low-emission vehicle (such as a Fiat 500) traveling at a steady 60 miles per hour and a fuel efficiency of 43 miles per gallon, the carbon comes out at 86 g CO_2e per person mile.

At the high end of the scale we have a single person in a poorly maintained, rapidly depreciating, high-emissions car that looks more like a tank, cruising at 90 miles per hour or driving unsympathetically in urban conditions with heavy use of both brakes and accelerator. In these conditions, a vehicle of this type may achieve as little as 7.5 miles per gallon.

My numbers are higher than those you normally see for driving. That is partly because I am including the emissions from the extraction, refining, and transportation of fuel, as well as just the burning of it. Even more importantly, I am factoring in the manufacture and maintenance of the vehicle itself.

As a rule of thumb, about half of the carbon impact of car travel comes out of the exhaust pipe itself.[32] A few percent come from the processes of extracting, shipping, refining, and distributing the fuel (see Gasoline, page 89). The rest, typically 40 percent of the footprint, is associated with the manufacture and maintenance of the car. Big, expensive new cars have more of their embodied emissions attributable to each mile of driving. An older car that is still fairly efficient could beat a new Fiat 500 by virtue of having had its embodied footprint written off. (See New car, page 141.)

But it's not just what model you drive that matters. Here are 10 good ways to reduce the carbon footprint of your car use:

> Use the train, bus, or bike if traveling alone. Typical savings: 40 to 98 percent. (See New York City to Niagara Falls return, page 116.)
> Put more people in the car. This could make it better than train travel, provided that the others were otherwise going to drive separately. Typical savings: 50 to 80 percent.
> Join a car-sharing scheme.
> Drive a small, efficient car. Typical savings: 50 percent compared with the average car.
> Look after your car so that it will do 200,000 miles in its lifetime and it runs as efficiently as it can. Typical savings: 30 percent compared with the average. (See New car, page 141.)
> Accelerate and decelerate gently, avoiding braking where possible. Typical savings: up to 20 percent in urban conditions.
> Drive at 60 miles per hour on highways and freeways. Typical savings: 10 percent compared with 70 miles per hour.

> Keep the windows up when driving fast, and the air-conditioning off. Typical savings: 2 percent.
> Keep the tires at the right pressure. Typical savings: 1 percent.[33]
> Avoid rush hour. (See Congested car commute, page 107.)
> Drive safely. (See Car crash, page 139.)

Is it worth slowing down?

Although we might know that driving more slowly on highways is better for the planet, this concern often gets outweighed by our desire to get there on time. After all, time is money, right?

I'm going to take the case of someone driving on their own and assume that they value their time at $24 per hour. That's about the take-home pay per hour of someone earning $60,000 for a 37.5-hour working week plus a half-hour highway journey (around 35 miles) to the office each way. This is above average for the population as a whole but may be about typical for those who commute on highways.

I have assumed that the commuter in question drives a car that is capable of 33 miles per gallon at 70 miles per hour but 45 miles per gallon at 60 miles per hour. That is reasonable because, in highway conditions, the fuel consumption is roughly proportional to the square of the speed.[34] I'm also going to assume that this person hardly cares at all about their impact on climate change (we're going to look at financial costs and benefits only) and that the carbon cost of gasoline is 3 kg (6.6 lbs.) CO_2e per liter, or about 11 kg (25 lbs.) CO_2e per gallon (see page 89).

As Table 4.2 shows, in this scenario, the slower driver saves carbon but loses a bit of money. In the U.K., where fuel is more expensive, the driver breaks even by slowing down. Those less well off or with hungrier cars would be better off as well.

	70 mph	60 mph
Value of driver's time, per hour	$24	$24
Miles per gallon	33 mpg	45 mpg
Carbon footprint of a 70 mile round-trip commute	21 kg (46 lbs.) CO_2e	16 kg (35 lbs.) CO_2e
Time cost of the 70 mile commute	$24	$29
Cost of gasoline	$6	$4
TOTAL COST	$30	$33

TABLE 4.2: How to save carbon without losing money.

Electric and hybrid cars deserve a mention. Let's be clear. The electricity has to come from somewhere. Until such time as we have abundant renewable electricity, any additional demand for the stuff has to be met entirely through fossil fuel, rather than wind or hydro turbines whirring round faster. So, for the foreseeable future, switching from gas to electric doesn't get you away from having a 100 percent fossil fuel–powered car. However, in principle, it does give a small efficiency gain, because although there is energy lost in the process of generating the electricity from fossil fuel and transporting it to your car, once there, an electric car can do an efficient job of turning that electricity into mechanical energy. The best gas cars, by contrast, can turn only about 25 percent of its fuel energy into motion, with the rest dissipated as heat. Electric and hybrid cars can also offer regenerative braking, generating electricity as you brake. This makes little difference on the highway but can be significant for urban driving. Overall, for the next few decades at least, electric cars stand to give us a few percent efficiency improvement but definitely not a revolution in the carbon friendliness of motoring. (See also New car, page 141.)

A red rose

Zero CO_2e picked from your garden, no inorganic fertilizer used
350 g CO_2e grown in Colombia and flown by air
2.5 kg (5.5 lbs.) CO_2e grown in a heated greenhouse in
the Netherlands and then flown
> A single red rose could have the same impact on climate
change as about five kilos (11 pounds) of bananas.

Could the banana ever replace the rose on Valentine's Day? If you try this low-carbon alternative, please let me know how you fare (info@howbadarebananas.com).

The numbers here sum up the Hobson's choice that you are faced with if you want out-of-season cut flowers. You either have to put them on a plane or grow them using artificial heat. Both of these are bad news for climate change.

The study I based my numbers on found that for consumers in the U.K., Dutch roses had about six times the carbon footprint of the air-freighted ones.[35] After all, Holland is a cold country in winter and roses take a long time to grow. This only adds up commercially because the Netherlands subsidizes the energy required by its floral industry. In the U.K., home-grown flowers will probably have enjoyed only the Sun's heat.

In my work on U.K. supermarket products, out-of-season cut flowers emerged as some of the products with the largest carbon footprint per pound generated at the tills. In other words they are one of the most carbon-unfriendly ways of getting rid of your cash.

There's another concern, too. All commercial cut flowers use land that could otherwise be growing food. The demand for agricultural land is already driving deforestation (see page 151), which in turn is responsible for around 18 percent of man-made emissions. Looked at in those terms, cut flowers have to mean less rainforest—so the true footprint is probably even bigger than my numbers suggest.

Quite a few people I've spoken to have said that their attraction to cut flowers wilted once they made the connection with the huge emissions and pressure on land that they bring about.

So, stick to your own grown garden crop if you can, and do without flowers when they are not in season. As for alternatives, longer-life indoor plants are a dramatically less carbon-intensive option. And some artificial flowers are just about indistinguishable from the real thing—if you can bear the concept.

1 kg (2.2 lbs.) of boiled potatoes

620 g (1.4 lbs.) CO_2e locally grown, boiled gently with the lid on
1,170 g (2.6 lbs.) CO_2e still local but boiled furiously with the lid off

This panful of potatoes contains two-thirds of a woman's daily calorific needs. If potatoes were all you ate for a year, you could feed yourself for just 330 kg (730 lbs.) CO_2e, or 3 percent of the 10-ton lifestyle. That is good going when you consider that food and cooking currently account for 3 tons CO_2e per person per year. (That's without taking account of deforestation, which could add half as much again). You'd end up bored and malnourished if you stuck rigidly to this regime, of course, but there is clearly a place for potatoes in the low-carbon lifestyle. Table 4.3 shows how the footprint breaks down.

	Grams CO_2e
Growing the potatoes	220
Transport	80
Packaging in a simple bag	10
Supermarket storage and display	60
Boiling	250 to 800
TOTAL	620 to 1,170 (1.4 to 2.6 lbs.)

TABLE 4.3: Breaking down the potato footprint.

Potatoes are a low-carbon crop; larger conventional varieties are especially so, simply because yields are higher.

Transport emissions are not high, provided these potatoes stay in the locality. It is not uncommon for some supermarkets to move produce hundreds of miles to a distribution center and then back again. Even when this happens, however, the transport does not have a disastrous impact.

The biggest part of the footprint comes from the cooking process. The way you do this can alter the total footprint by a factor of two. Here are some ways to keep the cooking emissions to a minimum:

> Use a gas stove.
> Use a lid on the pan.
> Boil gently. The temperature of the water, and therefore the cooking speed, is exactly the same when you turn the gas down to a gentle simmer as when you boil at full throttle.
> Cut the potatoes into smaller pieces.
> Use a pressure cooker: the pressure raises the boiling temperature, which means the potatoes cook faster and more efficiently.

Alternatively, if you are baking or roasting, you can:
> Use a microwave or a convection oven.
> Reduce the size of the pieces.
> Having heated the oven up, cook more than one thing.

I have ignored the carbon cost of getting to the store (see Driving 1 mile, page 66, and Cycling a mile, page 26).

A pint of milk

723 g (1.6 lbs.) CO_2e

> So if you get through two pints a day in your household, that's 527 kg (1,160 lbs.) per year, as much as a return flight from San Francisco to Vancouver.

Milk is high-carbon stuff for exactly the same reasons that beef is. Cows, like most animals, waste a lot of the energy in the food they eat in the process of simply keeping warm and walking around rather than creating meat and milk. In addition, cows ruminate (chew the cud), which means they burp up methane, roughly doubling the footprint of the food they produce.[36]

As Figure 4.7 shows, around 85 percent of the milk's footprint is generated on the farm, but transport, packaging, and refrigeration also play their part. Because milk is heavy, keeping it local (and not trucking it hundreds of miles to and from distribution centers) seems like a good idea. My instinct is that milk delivery services probably cut carbon footprints by keeping the weight of our shopping bags down and therefore making it that much easier to walk to the store for everything else. In addition, reusable glass bottles almost certainly beat plastic disposables, even if you recycle the latter every time.

Wherever you get your milk, however, it remains—like all food from cattle—a high-carbon way to get your calories. There is probably quite a lot that could be done to reduce its carbon cost, but it's a hugely complicated area to research. Various studies have been carried out so far, but they don't always agree. To give a flavor of how confusing everything is, if you change the feed, you alter the carbon cost of that feed, the milk yield, *and* the amount of methane that gets belched out. At the same time you play about with factors like the life expectancy of the cow, the amount of saleable meat that the herd will produce alongside its milk, and the other inputs that will be required to keep the cow healthy. To make things even more complex, different farming practices affect the ability of the soil to absorb and store carbon. And everything also depends on the location of the farm and the breed of cow. Nobody has yet properly worked out how all these variables interact.

If the carbon footprint were the only consideration, the unpleasant truth is that the most efficient thing to do would probably be to keep cattle in small indoor spaces and rear them as intensively as possible, minimizing wasteful activities such as getting exercise or keeping warm. But carbon isn't the only consideration, of course, especially for organic farmers such as David Finlay in southwest Scotland. David is reducing the milk output of his herd in southwest Scotland from 7,500 liters per cow per year to 5,000 (1,980 gallons to 1,320 gallons). He believes that although the milk yield will go down, the amount of meat he can sell will go up and his feed costs will fall, along with his use of antibiotics and other inputs that have both financial and practical costs. Hugely important to David are two other factors: the animals will have even better lives than they already have on his farm, and he stands to have more free time because he will only have to milk them once a day. He believes a system like this could compete in the supermarkets alongside conventionally farmed milk even without the organic label. In fact, he

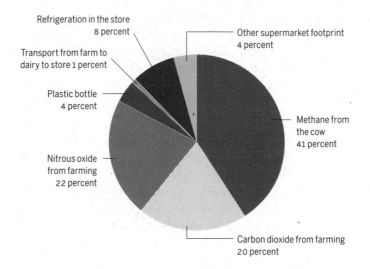

FIGURE 4.7: The carbon footprint of locally sourced milk in a plastic bottle at the checkout of Booths Supermarkets. In this example, the milk comes from Bowland Fresh, a local supplier, so the transport impact is low. This chart doesn't include either your journey to the shops or home refrigeration.

believes the price premium on the organic label comes mainly from the administrative costs of demonstrating at every step of the journey from farm to shop that no contamination with conventional milk has taken place.

One partial solution to the belching problem, legal in the U.S. since 2004 and widely used, is Rumensin, a simple additive that markedly cuts methane production in cows. The EU classified it as an antibiotic and banned it, even though the farmers I've spoken to say this was a mistake because it does not have the human health impacts usually associated with antibiotics. I'm not an expert on these things, but I can believe this might possibly have been a bureaucratic blunder that is now waiting to be overturned.

Whatever the truth about different dairy farming practices, soy milk is almost certainly a lower-carbon option than anything from a cow. Even though I haven't seen a study of this, in comparison with cows' milk, there is none of the inefficiency of putting animals in the food chain and no rumination involved. The market for soy is driving deforestation, but the problem is not the stuff that is eaten directly by humans: most soy is fed to . . . cows.

1 kg (2.2 lbs.) of cement

100 g CO_2e Eco-Cement
710 g (1.6 lbs.) CO_2e standard cement, efficient production
910 g (2 lbs.) CO_2e global average
1 kg (2.2 lbs.) CO_2e inefficient production

The world produces around 2.2 billion tons of cement per year—or around 300 kg (660 lbs.) per person. Nearly half of this (47 percent) is produced in China. Making this basic building material results in a staggering amount of CO_2e: around 4 percent of the world's total greenhouse gas footprint.[37] This figure is so high because the chemical process that turns limestone into cement gives off large volumes of CO_2 directly *and* takes a huge amount of energy.

Around half the footprint is down to the chemical reaction. There is not much you can do to reduce this without changing the product itself. About 40 percent comes from the burning of fuel to drive the reaction, leaving 10 percent for other bits and bobs in the cement industry and its supply chains.

Because of the basic chemical reaction required to make the stuff, it is hard to see how conventional Portland cement could be made into a low-carbon product. One alternative is Eco-Cement, a product invented by John Harrison in Tasmania. Eco-Cement's advocates claim not only that this product requires half the energy input of conventional cement but also that it reabsorbs CO_2 from the air as it hardens (around 400 g CO_2e per kilo). There are also claims that it is easier to incorporate waste materials into the mix than with normal cement and that it is easier to recycle. The product is based on magnesite, which is not as abundant as limestone, and perhaps that's why not everyone is using it yet. Or perhaps it is no good at sticking things together. I haven't tried it.

Cement makes up about 12 percent of the footprint of the U.K. construction industry, so other potential ways of reducing its impact are to use different materials, to build to last and build less, and to refurbish in preference to knocking down and building anew (see House, page 147).

1 kilo to 10 kilos
(2.2 pounds to 22 pounds)

A paperback book

400 g CO_2e recycled paper, with every copy printed getting sold
1 kg (2.2 pounds) CO_2e average
2 kg (4.4 pounds) CO_2e the same book on thick virgin paper,
with half the copies getting pulped
> The carbon footprint of a typical paperback is about the same
as watching 12 hours of programs on an average TV.

Overall, reading is a low-carbon activity, and there is plenty
of room for it in the sustainable lifestyle.* Why? It's hard to
drive or shop while you read. For a short while, a gripping
novel halts the consumerist lifestyle in its tracks.

My average figure is based on a 250 g book printed on
paper from a mix of virgin and recycled pulp.[1] I've assumed
that 60 percent of all copies made are actually sold, even
though I've heard more pessimistic estimates than this. The
economies of scale in printing are such that it pays to print
too many.

* "Sustainable lifestyle": this is a tricky expression. It doesn't bear much
scrutiny, and we could get hopelessly bogged down defining it. However,
I strongly suspect that whatever your definition I would still stand by my
assertion that it leaves plenty of scope for reading.

At the high end, the same book is printed on heavyweight high-gloss virgin paper and weighs 350 g. Half of the print run is pulped without ever hitting the shops.

At the low end, the book still weighs 250 g but is printed entirely on recycled paper. Roughly speaking, it takes about twice as much energy to make paper from trees as it does from recycled pulp—though the actual value varies enormously depending on the efficiency of the paper mill and the quality of the paper.

What you are reading right now doesn't yet exist as I write, but I'm guessing that, in carbon terms at least, you are holding a better-than-average paperback because my publisher thinks about these things. However, once you stop to think about it, there are all sorts of difficult questions about what to include in the sums. I haven't included the electricity burned by my computer as I'm typing right now, or any part of the footprint of my publisher's offices at Greystone, or a host of other possible elements.

Nonetheless, I hope this book pays for itself in carbon terms fairly easily. You have to cut out only about three car miles to cancel out its production.

All carbon footprints need to be thought of in terms of "bang for buck": do the benefits outweigh the impact? To maximize the "bang" side of the equation, you simply have to read this book, talk about it, and pass it around.

Electronic book readers deserve a mention. I guesstimate that an e-reader such as Kindle, Kobo, or iPad has a footprint of around 50 kg.[2] If I'm right, you'd have to get through at least a hundred paperbacks (bought new and then sent to recycling) before the paper savings outweighed the embodied emissions of the reader itself. This is before electricity consumption of the reader and in IT networks has been taken into account. E-readers may be wonderful devices, but I can't see a carbon argument for getting one, unless it gets you reading more. You can't yet take them in the bath, but potentially you can have your favorite e-bookshelf with you almost wherever you go.

A loaf of bread

1 kg (2.2 lbs.) CO_2e an 800 g (1.8 lb.) loaf

> Bread is good stuff: a year's calorific intake can be had for around half a ton CO_2e. That's only 5 percent of the 10-ton lifestyle and one-sixth of the current U.K. diet.

As Figure 5.1 shows, just over half the emissions of a loaf of bread come from the actual growing of the ingredients. About one-sixth is the baking. Transport is typically one-seventh, and the supermarket itself adds about one-ninth. The bag is a very small consideration—and if it helps to keep the bread fresh for longer, it is probably well worth it.

Bread is a great low-carbon food provided we actually eat it. There's the catch. It gets thrown away because we are fussy eaters and because it doesn't keep well. Tristan Stuart's eye-opening book *Waste* has a picture of a Marks & Spencer sandwich factory systematically discarding four slices from every loaf: the crust and the next slice from each end.[3] The

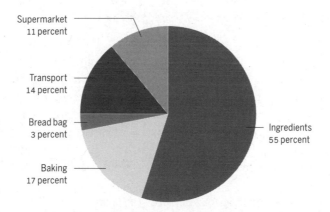

FIGURE 5.1: The footprint of bread at the supermarket checkout.

remaining slices get made into fresh sandwiches and are still at risk of being trashed before they are sold. Only once safely through the checkout do the odds of a sandwich being eaten start looking good, but there are still such hurdles as children who won't eat crusts and over-catered corporate lunches.

Loaves sold straight to consumers are no better, because the shelf life is so short. Plenty is trashed by the supermarkets, and plenty more goes stale in bread bins or ends up in a half-eaten sandwich. To keep the carbon cost of your bread to a minimum, buy only what you need, enjoy the crusts, and get your children to do the same. Find uses for stale bread: as toast, as an accompaniment to soup, and so on. And buy smaller loaves if you are not getting through the larger ones— the introduction of the 600 g loaf will help with this.

A bottle of wine

400 g CO_2e from a carton, with few road miles
1,040 g (2.3 lbs.) CO_2e average
1,500 g (3.3 lbs.) CO_2e over-elaborate bottles, transported for thousands of miles by road

> So if you drink three bottles of typical wine per week, which is pushing the limits of a healthy lifestyle, that is about 150 kg (330 lbs.) per year, equivalent to driving 210 miles in an average car.

My estimates here are based on a study I did for Booths supermarkets (Figure 5.2). For a typical bottle, just over one-third of the footprint comes from the production of the wine itself. Whether or not it is possible to reduce this by buying organic wine is not clear, although there may be other environmental benefits of the organic option. It is difficult to know from the label what the carbon intensity of a particular vineyard is, so I have just given all wine a typical value, based on various studies.

The glass bottle accounts for a similar amount of carbon to that of the wine it contains. There is a simple savings to be made here: by buying wine boxes or cartons, you can reduce the footprint of the packaging by a factor of about five. In doing so, you will also reduce the weight, so transport emissions can also be slashed by one-third. There will be absolutely no loss of quality, even though you might lose some choice. If the carton offends you, you can always decant the wine into a jug.

There is a lot that can be done without getting rid of the glass altogether. Organico is a wine distributor near where I live that has started importing some of its wine unbottled. This cuts the transport weight. It does its own corking and puts a £2 ($3) deposit on the bottles, which are themselves 15 percent lighter than normal and are made from clear glass because this is better for eventual recycling. One further nice touch is that they have done away with the concave bit under the bottle that has always struck me as fundamentally dishonest.

Note that shipping is only a small component, so it doesn't matter all that much what continent your wine comes from. Far more important are the road miles—both in your country and in the country of origin. For this reason, locally produced wine could cut the footprint by 25 percent, provided that your neighborhood has the right kind of climate. A

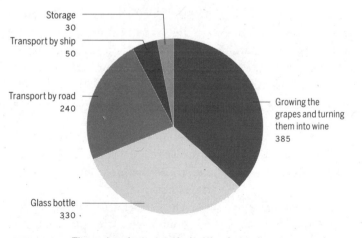

FIGURE 5.2: The carbon footprint of a bottle of wine (in grams CO_2e).

New Yorker is probably better off drinking wine shipped from Europe than by road from closer-by California.

Because it is less dilute, wine often turns out to be a slightly less carbon-intensive way of taking alcohol on board than beer (see page 50).

All these calculations assume that you recycle any packaging.

Cork, plastic, or screw top? In carbon terms all are insignificant (see *misdirection of attention*, page 8), but the good old-fashioned cork won't be in any danger of bobbing around in circles in the Pacific for a thousand years.

1 kg (2.2 lbs.) of plastic

0.75 kg (1.7 lbs.) CO_2e Ecosheet
1.7 kg (3.7 lbs.) CO_2e PET for plastic bottles, from recycled materials
3.4 kg (7.5 lbs.) CO_2e polystyrene from virgin materials
3.5 kg (7.7 lbs.) CO_2e average
4.4 kg (9.7 lbs.) CO_2e polypropylene for injection molding, made from virgin materials
9.1 kg (20 lbs.) CO_2e some types of nylon[4]

Plastic is such useful stuff: it's tough, durable, and waterproof. No wonder we use so much of it. Unfortunately, plastic tends to be so durable that it hangs around in landfill sites for centuries, clutters up the stomachs of animals and fish, transforms remote Scottish beaches into junkyards, and ends up in almost every ecosystem you can think of. But from a purely carbon perspective, its inability to rot is good news in as much as it won't add to methane emissions from landfill: if we assume that the plastic is put in the trash rather than tossed into a street or field, those hydrocarbons are going back underground where they came from.

As the figures above show, the footprint of making plastic from virgin material is about double what it would be if recycled products were being used. The challenge for recycling plastics is that it's difficult but necessary to separate the various types and process them separately. This isn't true for Ecosheet, however. This brand-new construction material can be made from the full range of different plastics, so almost nothing goes to landfill. Once you have finished with it, the sheeting can even be reworked into new boards. The makers, 2K Manufacturing, told me that they don't even need to heat up waste plastic to the usual recycling temperatures to create their boards. As I type, only a few sample boards per day are being produced, but by the time you read this, full production is expected to have been underway,[5] the Science Museum in London will have used the stuff to build its new exhibition on climate change, and it should be available in North America. Biodegradable plastic packaging is worth a mention because it can be a well-intentioned disaster area. It sounds great, but if you send it to landfill, it rots down and emits methane, and if you throw it into the recycling, it can ruin the entire batch. It should be compostable instead, but I have also heard that it releases chemicals that slow down the degrading process for the rest of the trash or heap.

Taking a bath

Zero CO_2e heated by solar energy
0.5 kg (1.1 lbs.) CO_2e modestly filled, efficient gas heater
1.1 kg (2.4 lbs.) CO_2e generously filled, efficient gas heater[6]
2.9 kg (6.4 lbs.) CO_2e generously filled, electric water heater
> A daily bath adds up to between 180 and 1,050 kg (between 396 lbs. and 2,310 lbs.) CO_2e per year—that's between 2 and 10 percent of the 10-ton lifestyle.

In our family at least three of us often end up using the same water, even if not all at the same time. (Anyone who's been running through mud has to go last.) Since we top up with hot, the bath is always full to the brim by the end. That is about 120 liters (31 gallons), giving a footprint per person of around 400 g.

If you were to read a book in the bath for an hour, you'd probably add 50 percent to the footprint of the average full bath by pulling out the plug with your toes from time to time and topping up with hot. So the actual leisure activity would be 500 g (1.1 lbs.) per hour on top of the functional bath itself. That is quite a bit higher than most TV watching but still a lot lower than any pastime that involves using a car.

In winter you can reclaim about half the heat simply by leaving the plug in until it goes cold. This works provided that you actually want the heat in your bathroom and don't object to the idea of old bathwater hanging around.

The comparison with showers (see page 53) might be a surprise. Electric showers on the market range from 7 kilowatts at the weedy end to 11 kilowatts at the powerful end. For the same impact as a full bath from an efficient gas furnace you could have a 9-minute high-volume "power shower" or a 13-minute weedy shower. In comparison with sharing bathwater, you would have to be a family of fast scrubbers to make the electric shower win out. That's even without taking account of any bathwater heat reclaimed (which isn't an option for the shower unless you have a plug and are prepared to stand in an ever-deepening pool). To be fair, though, showers where the hot water comes from a gas furnace, which is increasingly the norm, are much more carbon efficient and generally will work out as using less energy than a bath unless the latter is shared among many people.

What about other options? A trip to the swimming pool could have a much higher impact than a bath, even if you were to walk there (see page 150), whereas a wild swim comes out best of all—though not many of us live near a clean and safe river or lake.

Overall, baths do serve a purpose, and even the most luxurious needn't be too bad as long as they are not electrically heated. Then again, if everyone in your household has extravagant bathing habits, they could easily come to over 1 ton per year.

A bunch of asparagus

125 g CO$_2$e a 250 g pack, local and seasonal
1.9 kg (4.2 lbs.) CO$_2$e the same pack, air-freighted from
Peru to the New York in January
3.5 kg (7.7 lbs.) CO$_2$e the same pack, air-freighted from
Peru to the U.K. in January

> If you live in New York and your entire diet were as carbon intensive as long-haul asparagus, your food footprint alone would be more than the entire footprint of the average North American. If a Londoner did the same, the footprint of his or her food alone would be more than three times the average U.K. citizen's total footprint.

The numbers here are based on data from Booths supermarkets, which to their credit took steps to increase their local sourcing when they saw the impact of the Peruvian product and are now emphasizing the benefits of seasonal food more strongly than ever. Air-freighted from Peru to New York, asparagus comes in at 8 kg CO$_2$e per kilo (3.6 kg CO$_2$e per pound) or, to put it another way, about 40 g of carbon per calorie. It is over 50 times more carbon efficient to get your calories from bread.

When produce is being moved, a mile by air has more than 100 times the climate impact of a mile by sea. This is because it takes a lot of energy to keep a plane in the air—and also because engine emissions tend to do more damage at high altitude than they do at ground level (see Flying from Los Angeles to Barcelona, page 133). For this reason it is difficult to see how there can be any place at all for air-freighted food in a sustainable world.

Examples of other foods that are very likely, when out of season, to have been air-freighted or (just as bad) grown in an artificially heated greenhouse include baby corn, baby carrots, snap peas, small green beans,[7] fine beans, okra, shelled peas, lettuces, blueberries, raspberries, and strawberries.

At the other end of the scale is asparagus grown in season in your own country. This cuts out a staggering 97 percent of the footprint. When asparagus is out of season (which is most of the time), try to favor low-carbon options such as kale, carrots, parsnips, turnips, or leeks.

Flying from closer-by California, Washington, Michigan, or even Mexico has less impact than flying from Peru. And at each end of the local asparagus season there are periods in which a small amount of heating makes the crop viable.

None of the estimates here include the footprint of cooking the food, which is likely to be around 100 g CO_2e if you simmer it for 8 minutes with the lid off.

A final comment: the recipe book I consulted advised strongly against air freight on taste grounds, stressing the importance of eating asparagus within 48 hours of harvesting.

A load of laundry

0.6 kg CO_2e washed at 30°C (86°F), dried on the line
0.7 kg CO_2e washed at 40°C (104°F), dried on the line
2.4 kg CO_2e washed at 40°C (104°F), tumble-dried in a vented drier
3.3 kg CO_2e washed at 60°C (140°F), dried in a combined washer-drier

> If you wash and dry a load every two days, that's equivalent to 440 kg (970 lbs.) CO_2e, which is like flying from London to Glasgow and back with 15-mile taxi rides to and from the airports.

Modern washing powders work just as well at 30°C (86°F), so there is a very simple savings to be had here of 100 g CO_2e per wash just by turning the temperature down. But the bigger savings relate to drying. As the numbers above show, for a typical 40°C (104°F) wash nearly three-quarters of the carbon footprint comes from the drying rather than the washing. Tumble driers generally use electricity to generate heat. This is more than twice as carbon intensive as generating heat from gas. If you use a conventional vented drier, most of the heat is simply pumped out to the outside world, so overall it's a wasteful activity. Condensing driers use a little bit more energy still, although, in winter at least, all that heat stays inside your house, where it is probably of some use.

Overall, a household running the tumble drier 200 times a year could save nearly half a ton by installing a clothes rack inside and a washing line outside. In winter the evaporation from drying clothes will cool your house down slightly, but it's a marginal effect, and on a baking hot summer's day, our clothes drying in the kitchen act as free air-conditioning.

Make sure that your washer has a good spin function. It is much more efficient to remove most of the water by spinning it off than by evaporating it.

All the figures listed above are based on a full 5 kg (11 lb.) load (half loads use a little less energy each time, but they work out as much less efficient per garment washed). I've allowed around 220 g per wash for the embodied emissions in the appliances themselves.[8] If this estimate is correct, the manufacture and delivery of the appliances account for nearly 10 percent of the total carbon footprint of each wash.

You can probably improve on the lifetime of your washer and/or drier if you look after it and get it repaired when it breaks. Switching from a typical 1998 machine to a new one with an "A" rating might gain you around 10 percent in efficiency[9]; in other words, roughly enough to offset the emissions required to make the new machine but no more. So the message is that unless your machine is particularly cranky and inefficient, there is no real carbon case for getting a new one.

While on the subject of washing, have you optimized the frequency with which you wash stuff? I don't want you to start going around smelly, but it's worth asking the question: does stuff go in the wash unnecessarily often? If you can reduce the number of washes you do without upsetting anyone, there is a time savings to be had, too, so it's a great example of life getting better as the carbon comes down.

A burger

1 kg (2.2 lbs.) CO$_2$e veggie burger
2.5 kg (5.5 lbs.) CO$_2$e 4-ounce cheeseburger

> If you eat a cheeseburger each day, that's a massive 910 kg (2,000 lbs.) CO$_2$e per year—the same as driving 1,500 miles in a fairly efficient car and just over 1 month's worth of ration in the 10-ton lifestyle.

The 4-ounce cheeseburger described here provides 515 calories. If this were the only type of food you ate, the average man would need about five burgers per day provided he didn't do much exercise. (The average woman could get away with one fewer.)[10] If you managed to keep up this diet for a year without killing yourself, you'd cause about 4.6 tons of carbon emissions just through your food.

The cheeseburger's footprint breaks down as shown in Table 5.1.

The biggest factors here are the beef and the cheese. As we've already seen, animal products tend to be more carbon intensive than vegetables and grains because animals consume a lot of energy just to keep themselves warm and move around. This makes their job of converting animal feed into meat and milk inherently inefficient.

There is another big problem with beef and dairy farming. Cows, like sheep, are ruminants. This means that they belch

Component	Grams CO$_2$e
Beef (108 g)	1,910 (4.2 lbs.)
Cheese (20 g)	250
Bun (40 g)	50
Salad (20 g)	10
Condiments (20 g)	80
Cooking and transport	200 (approx.)
TOTAL	2,500 (5.5 lbs.)

TABLE 5.1: The carbon footprint of a 4-ounce cheeseburger.

out methane, a greenhouse gas 25 times more potent than CO_2. The result is that beef and lamb have around double the carbon footprint per kilogram of meat compared with that from pigs.

A further consideration is that excessive demand for meat provides an incentive for deforestation because it raises the demand for arable and grazing lands. That said, there is plenty of land, for example in the U.K., that is fit for cattle and sheep farming but not for crops, and there can be a conservation benefit in having animals on the land.

It is unclear whether the footprint of the burger could be reduced by using organic or free-range meat and cheese. The Cranfield University study[11] I have used for my figures found that organic cattle farming had few or sometimes negative carbon benefits. However, the organic farmers I know who have studied this report were scathing of the assumptions made about organic practices and yields. The carbon benefits of rough grazing are also unclear. On the one hand less feed is required, but on the other hand there are complex implications on yield and rumination.

At the time of writing, my inclination is to say that a low-carbon diet can safely contain a bit of beef and lamb from rough pastures, but the whole area clearly warrants further research. And in the meantime, there's always the low-carbon veggie burger option.

See also Steak (page 97) and Deforestation (page 151).

A quart of gasoline

3.15 kg (6.9 lbs.) CO_2e

> In the U.K. we get through about 50 billion liters (13.2 billion gallons) of gas and diesel per year. That's something like four bathfuls, right up to the brim, for every man, woman, and child in the land.

The pie chart in Figure 5.3 speaks for itself. If you were to pour a quart of gasoline on the floor and strike a match, the fumes

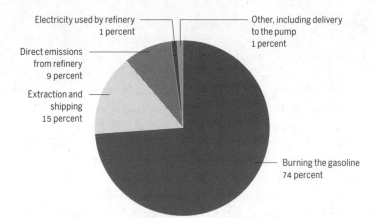

FIGURE 5.3: The footprint of gasoline is more than just the CO_2 that comes off the fuel.

would account for only about three-quarters of the carbon story. The other quarter is caused by the supply chain of the fuel: getting it out of the ground, flaring off the gas, shipping it around the world, refining it, and getting it to the pump.

This extra quarter doesn't usually feature in car emissions statistics (including official greenhouse gas "conversion factors"), which generally deal only with the stuff that comes out of the exhaust pipe. This is one part of the reason why the carbon footprint of driving is often so badly underestimated.

The story for diesel is slightly different. Each quart has a slightly higher footprint (13 percent), but it has a proportionately higher energy content to compensate. Diesel engines are typically about 30 percent more efficient at turning fuel energy into vehicle movement. But if only it were that simple. The downside is that, although they have got much cleaner in recent years, diesels also belch out many more particulates, and this also contributes to climate change (see Black carbon, page 169) as well as causing asthma. Overall, it is hard to say which wins as the environmental vehicle fuel.

Biodiesel deserves a mention as a controversial option, full of problems but also with some potential for the future. The first thing to say is that using land to grow fuel rather than food puts pressure on the world's forests, and chopping trees down already accounts for 18 percent of global emissions (see Deforestation, page 151). With a fast-growing world population, land is likely to feel increasingly scarce in future. The second negative point to make is that considerable emissions are involved in the growing of the fuel crop and the process of turning it into fuel. Some people even think this can even add up to more than the emissions from fossil fuels. On the plus side for biofuels is the potential to create them from unavoidable agricultural waste and the prospect that future technologies may be able to create them efficiently from algae. Overall, biofuels might one day be a useful part of the solution, but they are not likely to be a magic wand.

1 kg (2.2 lbs.) of rice

2.5 kg (5.5 lbs.) CO_2e efficiently produced
4 kg (8.8 lbs.) CO_2e average
6.1 kg (13.4 lbs.) CO_2e inefficient production with
excessive use of nitrogen fertilizer
> A typical kilo (2.2. lbs.) of rice causes more emissions
than burning a quart of diesel.

Rice deserves a place in your consciousness not only as a food on your table but as an important piece of the global jigsaw. It provides 20 percent of the world's food energy in exchange for 3.5 percent[12] of its carbon footprint (Figure 5.4). Its global footprint is just a fraction less than that of cement production (see page 75). Europeans and Americans get just 1 or 2 percent of their food energy from this crop, but the figure is very much higher in Asia, where 89 percent of the world's total rice harvest is consumed.[13]

I suspect that plenty of greens will be slightly unsettled to hear that rice, the simplest of foods, is a surprisingly high-carbon staple, much more so than wheat, which is nutritionally similar. That's because of the methane that bubbles out of

FIGURE 5.4: The carbon footprint of 1 kg (2.2 lbs.) of rice (in kg CO_2e).

the flooded paddy fields and the excessive helpings of fertilizer that are all too often applied.

Around the world, 600 million tons CO_2e of methane is thought to be emitted from rice paddies, accounting for around 1.2 percent of the total global footprint and about three times the footprint of all the cement produced in Europe. Even more significant are the 161 million tons of fertilizer, mainly nitrogen based, that are applied to the crop.[14] That's a little over 1 ton of fertilizer for every 3 tons of rice produced. If this is made in an efficient factory and applied sparingly, at well-chosen moments, each ton applied may only result in 2.7 tons CO_2e. If not, the figure could be as high as 12.3 tons CO_2e.

I have guesstimated just 100 g CO_2e per kilo (50 g per pound) of rice for the production of agricultural machinery and the transport of the rice to market. The majority is eaten in its country of origin, and I can't imagine rice ever finding its way into a hothouse or an airplane.

If we were to feed the world entirely on food that was as carbon intensive as rice per calorie, what would happen?

With smart use of fertilizer, global emissions from agriculture would come in at 11 percent of the current total, compared with the current 18 to 20 percent. But if the worst practice were the norm, agricultural emissions would increase.[15]

It's possible to grow rice without flooding the field and thereby cut out the methane. However, it's harder work (because you have to do more weeding) and you might need more fertilizer, which would mean a trade-off that could end up taking the net carbon impact either way. As with lots of agriculture, we don't fully understand what goes on or what the best options are. It is another important area for research, as the number of human mouths soars ever higher, along with the global temperature.

See also Fertilizer, page 136.

Desalinating a cubic meter (260 gallons) of water

Zero CO_2e solar powered (technology permitting)
5 kg (11 lbs.) CO_2e average, a reverse osmosis plant
in Sydney using electricity from coal
23 kg (51 lbs.) CO_2e inefficient thermal desalination plant
> Globally, desalination accounts for perhaps 0.6 percent
of our footprint—and this is rising fast.

Each day the world desalinates around 60 billion liters (16 billion gallons) of water—that's 60 million tons—and that figure is currently doubling about every decade.[16] Something like half of the total takes place in and around the dry, oil-rich Middle East, but desalination also accounts, for example, for 13 percent of California's electricity usage and 31 percent of its gas consumption. Emissions per liter vary hugely depending on the efficiency of the process and the carbon intensity of the electricity used. If a new plant in Sydney is typical of the global efficiency (it uses relatively efficient technology but powers it with electricity from coal), that leaves a global carbon footprint of about 300 million tons CO_2e—or something like 0.6 percent of all global emissions. And that figure is likely to continue increasing rapidly, not least because the world is getting hotter and drier in many regions—a feedback loop of climate change. Spain, which looks set to be one of the countries within Europe most directly affected by changing climate so far,[17] doubled its desalination between 2000 and 2004.

At the high end of the spectrum are inefficient thermal desalination processes. A big improvement on this is reverse osmosis. Various options exist for using spare heat from fossil fuel–burning power stations (though we should be careful not to double-count the benefits: in these situations the desalination plant may claim carbon neutrality while the power station claims to be offsetting its emissions by supporting the desalination plant).

At the low-carbon end, Seawater Greenhouse[18] claims to have developed a technique for using solar heat to desalinate water for greenhouse-cultivated crops in arid regions. In theory the desalination itself is just about carbon neutral. I haven't personally investigated the technique, but the company has won awards and has some large pilot projects already in operation. I include it here because it is the kind of technology that gives hope in the midst of increasing desertification problems around the world.

Apart from greenhouse gases, another nasty by-product of desalination is the brine concentrate that is returned to the sea, increasing the salinity and messing up marine ecosystems.

A pair of pants

3 kg (6.6 lbs.) CO_2e my favorite old nylon traveling pants
6 kg (13 lbs.) CO_2e my cotton jeans
> "Natural" materials may sound greener, but the footprint tells a different story.

My cotton jeans weigh 600 g (1.3 lbs.). Cotton has a footprint of around 7 kg CO_2e per kilo (3.2 kg CO_2e per pound).[19] On top of that there is dying, cutting, sewing, an allowance for waste fabric, buckles and zips, transport, and so on, which probably takes the total to about 6 kg (13 lbs.) per pair—equivalent to an 8-mile drive in an average car.

But this figure doesn't tell the whole story. Over the 4 years I suspect I'll get out of them, the footprint of washing and drying them is likely to be several times the footprint of producing them in the first place. My best estimate is four times more (Figure 5.5).[20] It all depends how quickly I get them dirty (quite quickly) and how tolerant I am of the dirt (probably more than most) and how they are washed and dried. But there's no avoiding the fact that the jeans are heavy when wet and they take a lot of drying.

At the other end of the scale is a pair of pants I've had for 12 years. They have traveled extensively, and I've worn them endlessly around my home, too. They weigh just 250 g, and they dry fantastically fast. I can't seem to wear them out. They cost $75, which felt like a lot at the time but now seems a bargain. They are made of some fancy fine-woven nylon. Nylon has a footprint of between 7 and 9 kg CO_2e per kilo (3.2 and 4.1 kg CO_2e per pound) depending on the specification[21]—so 12 kg CO_2e per kilo (5.5 kg CO_2e per pound) of finished pants, or 3 kg (6.6 lbs.) per pair, is probably about right.

Even if I wear my cotton jeans right into the ground, I can't envisage getting more than 200 days of solid use out of them. That works out to a minimum of 30 g CO_2e per day—or more than 100 g per day once I factor in the laundry. By comparison, my nylon pants are probably good for 600 days of wear, so that's just 5 g per wear-day, and because they virtually drip dry, the laundry aspect probably adds only an extra 6 to 12 g. All told, then, the nylon pants are less than one-tenth as carbon intensive as the jeans.

What about the rest of my clothes? If a pair of pants makes up one-quarter of my daily clothing, and if everything I wore were equivalent to my jeans, my clothing footprint would be 45 kg (99 lbs.) CO_2e per year for the garments themselves plus around the same again for washing and more than double

FIGURE 5.5: The annual carbon footprint of buying and washing clothes.

again if I use a dryer. But if everything I wore were equivalent to my nylon pants, I could cut my clothing footprint to just 7 kg (15 lbs.) per year, or 16 kg (35 lbs.) including laundry.

In reality, the average U.K. person has a clothing footprint closer to 225 kg (496 lbs.) per year, or more than 400 kg (882 lbs.) including laundry, which suggests, not surprisingly, that we are nothing like as good at wearing things into the ground as my scenarios suppose.

All told, if you live in the U.K., clothing and textiles will typically make up about 2 percent of your footprint. And there are broader environmental issues to consider, too. For example, the Aral Sea is drying up partly because of cotton plantations in its catchment[22] and the clothing and textiles industry produces toxins that find their way into water supplies.

Here are some tips for keeping the total impact of your clothing to a minimum:
> Buy stuff that is easy to wash and dry.
> Buy stuff that is built to last.
> Wear it and use it until it falls apart, or pass it on.
> Buy second-hand.

> Repair things rather than throwing them out.
> Donate or recycle clothing rather than putting it in the trash.
> Favor synthetic fibers over natural ones.

A steak

2 kg (4.4 lbs.) CO_2e a raw 4-ounce beefsteak

> A steak has about the same impact as 25 bananas. If you
> have one per day, that would be more than 700 kg (1,540 lbs.)
> CO_2e per year, equivalent to 1,000 average car miles.

Beef is a climate-unfriendly food, coming in at almost 18 kg
(40 lbs.) CO_2e per kilo (or 8 kg CO_2e per pound).[23] About nine-
tenths of this footprint comes from the beef farming itself. As
we've already seen, using animals to produce food tends to
be inefficient compared with eating crops (see page 47), and
cows have the added problem that they ruminate, producing enough methane to roughly double the climate change
impact of farming them.

Less widely discussed than the methane are the nitrous
oxide emissions, which account for about three-tenths of the
footprint of beef farming. This gas is released when nitrogen
fertilizer is applied to grass and other fodder crops, and when
the grass is silaged. Last, there is the CO_2 itself, at around
one-fifth of the farming footprint. This is caused by the trac-
tors, other farming machinery, and energy required to make
fertilizer.

I'm using the same footprint figure here for all beef,
although you could argue the case for attributing more to the
most expensive cuts than to the mechanically reclaimed stuff
that finds its way into economy burgers. In that sense, offal
and processed meat may well be a greener choice than more
premium meat products. But however you look at it, food
from cows remains toward the top of the carbon spectrum—
despite the ongoing debate about whether the footprint of
beef and milk can be reduced (see Milk, page 72).

A carton of 12 eggs

3.6 kg (7.9 lbs.) CO₂e

> So a single egg has about the same footprint as
  four bananas, even before you cook it.

If you tried living entirely off eggs (and survived the choles-
terol overdose) you'd meet your calorific needs for around
30 percent of the total footprint of the 10-ton lifestyle. This
makes eggs less carbon intensive than some animal products but more so than most vegetable-based foods. Figure 5.6
shows that, as with nearly all foods, most of the impact of egg
production comes from the farming itself (in this case the
rearing of birds and growing of their feed) rather than the
packaging or transport.[24] Chickens don't ruminate, so methane isn't much of a problem. But nitrous oxide is the main
contributor to the footprint of the final product.

I've based my figures for the farming part of the footprint on a study by Cranfield University. This study suggests
that—from the perspective of climate change—organic eggs
come out about 25 percent worse than those from battery

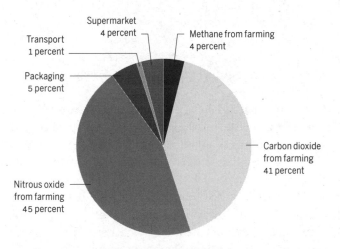

FIGURE 5.6: How the carbon footprint of eggs (not including cooking)
cracks up.

farms. This just goes to illustrate that if responding to climate change sends us into a blinkered drive for efficient production, some other values are going to have to suffer. This book isn't about telling you what values to have, but from time to time it's worth remembering that climate change is not the only issue. If you care about animal welfare as well as climate change, buying fewer eggs but making them organic might be a sensible compromise.

1 kg (2.2 lbs.) of tomatoes

0.4 kg CO₂e organic loose tomatoes, traditional variety, grown locally in July
9.1 kg (20 lbs.) CO₂e average
50 kg (110 lbs.) CO₂e organic "on the vine" cherry tomatoes, grown in Ohio in March
> For the same impact as just 1 kg (2.2 lbs.) of the most energy-intensive tomatoes, a man weighing 224 pounds could eat his own body weight in oranges.

Shocking! Tomatoes, at their worst, are the highest-carbon food in this book. But at their best, tomatoes are perfectly fine.

At the low end of the scale a high-yield classic variety is grown in the summer with no artificial heat required.[25] The middle and high figures are based on a detailed but controversial study by Cranfield University.[26] The middle figure is averaged across all varieties and times of year. Classic loose tomatoes, the ones that our parents were brought up on, cause only just over half the carbon, at 5.6 kg CO₂e per kilo (2.8 kg CO₂e per pound), whereas "specialist" varieties (defined here as cherry, plum, cocktail, beefsteak, and others) come in at almost 30 kg CO₂e per kilo (13.6 kg CO₂e per pound) because the yield is so much lower per acre, so they need more heat per pound.

Perhaps Cranfield's most unsettling finding was that when heat from fossil fuels is required, organic is the highest-carbon option, again because the yield was thought to be lower. At the high-carbon end, therefore, with a staggering 50 kg of greenhouse gas per kilogram (23 kg CO_2e per pound), are out-of-season organic, cherry tomatoes sold "on the vine."

So, tomato lovers concerned about climate change would do well to stick to the season (July to October in most of North America) and to favor classic varieties, sold loose. In the winter, it makes carbon sense to stick to canned tomatoes, but if you do want to buy fresh tomatoes outside the local growing season, it is almost certainly preferable to buy them from Mexico, California, or another warmer place rather than choose local versions produced in heated greenhouses.

1 kg (2.2 lbs.) of trout

5.9 kg (13 lbs.) CO_2e canned
6.9 kg (15 lbs.) CO_2e fresh from the store or frozen

I've chosen trout for my example even though I'm not sure I've ever seen it in a can. That's because up to the point of slaughter it has clocked up a carbon footprint that's fairly average for fish. At this level, fish comes out as a carbon improvement on beef and lamb—and it's healthy, too. But before you rush out and switch your diet over to eating the stuff seven days a week, bear in mind that many of our fish stocks are getting dangerously depleted, and if we all switched over from ruminant meat we'd probably wipe out global fish stocks in a decade.

I hesitate to mention that in the studies I looked at, overfished cod comes out with a slightly lower carbon footprint than salmon.[27] Once again, then, we each have to balance up carbon emissions with other concerns. The list in Table 5.2 will help shed some light on the carbon part of the puzzle. For the fish-stocks part, check out SeaChoice's handy fish-purchasing guide at www.seachoice.org.

When canned fish are compared with fresh stuff over the counter, the refrigeration is a bigger deal than the can, so fresh fish come out slightly higher in carbon (Figure 5.7). But then the fresh version is 100 percent fish, with no added oil,

Fish	Carbon footprint	
	(kg CO_2e per kilo)	(kg CO_2e per pound)
Mackerel, fresh fillet (at the factory gate)	0.5	0.2
Herring, fresh fillet (at the factory gate)	1.3	0.6
Unprocessed shellfish (freshly landed)	2.6	1.2
Unprocessed general fish (freshly landed)	2.6	1.2
Cod, fresh fillet (at the factory gate)	2.7	1.2
Cod, frozen fillet (at the factory gate)	2.8	1.3
Cod, fresh fillet (at the checkout)	2.8	1.3
Cod, fresh fillet (at the factory gate)	3.0	1.4
Cod, frozen (at the checkout)	3.2	1.5
Trout, frozen fillet (on leaving the slaughterhouse)	4.1	1.9
Flatfish, fresh fillet (at the factory gate)	4.7	2.1
Frozen filleted fish	6.5	3.0
Shrimp, peeled and frozen (at the factory gate)	10.1	4.6

TABLE 5.2: The carbon footprint of different seafoods according to various studies.[28]

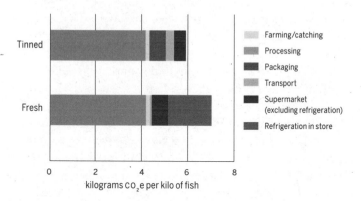

FIGURE 5.7: The carbon footprint of canned and fresh fish.

and it usually tastes better, too. Fresh fish has a similar footprint to frozen up to the time when you buy it. But the footprint of the frozen version could carry on going up for months in your freezer if you forget it is there.

Leaving the lights on

90 kg (198 lbs.) CO_2e a low-energy bulb for 1 year
500 kg (1,100 lbs.) CO_2e a 100-watt incandescent bulb for 1 year

By "incandescent bulb" I mean the old fashioned kind with the glowing tungsten wire. In the U.K. it is now illegal for stores to buy them, so they will soon be museum pieces—at last. Leaving a light on for a whole year might sound extreme, but having an average of one bulb turned on unnecessarily at any one time is almost certainly quite common. (In my office building, which is just 3 years old, and oddly rated environmentally "Excellent" by the Building Research Establishment, the light is permanently on in the shower. There is simply no off switch.)

As the figures above show, low-energy bulbs have the potential to save an enormous amount of electricity. However, efficiency alone won't bring about a low-carbon world, because the less costly something becomes, the more we tend to use it—so the result can end up being more consumption. In the case of lighting, this translates to "I've left a few lights on, but it's OK because they are low-energy ones." There's also the fact that the money we spend on bills will end up being spent elsewhere—a cheap flight, perhaps (see Rebound effect, discussed within The world's data centers, page 160).

Like any form of electricity wastage, leaving the lights on is one of the cheapest ways of trashing the planet, though the precise impact depends on where you live. I've based the figures here on a typical U.K. energy mix, but if you live in Australia, the footprint is about 60 percent higher. Some people might argue that if you live in France, it's OK to leave the lights on because it mostly comes from low-carbon nuclear power, but in my analysis that doesn't stack up (see Unit of electricity, page 56).

Finally, there's no truth in stories you may have heard that the act of turning a light on uses the same energy as leaving it on for half an hour.

1 kg (2.2 lbs.) of steel[29]

0.42 kg (0.92 lbs.) CO_2e recycled general steel
2.75 kg (6 lbs.) CO_2e virgin general steel
6.15 kg (13.6 lbs.) CO_2e virgin stainless steel
> So virgin general steel has about three times the carbon footprint of cement (or, for that matter, porridge made from half milk and half water) per pound.

These figures are for steel in its raw form at the foundry gate. In other words they do not take account of any additional emissions that might be required to transport it to wherever it will be used or turning it into something useful like a car or a part of a house. The value of recycling immediately becomes evident, because recycled steel has less than one-sixth of the footprint of its virgin equivalent.

Another key factor is the country of manufacture. This can make a big difference, for three reasons:
> Steel manufacture requires electricity, and the carbon intensity of this varies from country to country depending on how it is generated.
> The amount of energy used depends on the efficiency of the steel plant, with less developed countries often having less efficient manufacturing.
> A final small consideration is that if the steel is made a long way from its final market, there will be an additional shipping impact.

A report for the U.K. government[30] estimated that the emissions associated with the manufacture of 1 ton of steel in China were typically three times those for steel made in the U.S., where production is carried only two-thirds of the footprint per ton of steel made in Denmark or the U.K. only half (Table 5.3). India came out worse than China, and Nigeria is worse still—at over 11 times more carbon-intensive than for U.S. production.

Country	Carbon emissions % of U.S. production	Country	Carbon emissions % of U.S. production
Australia	125	Korea (South)	75
Austria	225	New Zealand	450
Brazil	175	Nigeria	1125
Canada	200	Norway	550
China	300	Pakistan	326
Colombia	600	Poland	225
Czech Republic	275	Russian Federation	350
Denmark	75	Slovakia	250
France	150	South Africa	475
Germany	200	Sweden	150
India	500	Ukraine	300
Indonesia	500	U.K.	150
Iran	100	U.S.	100
Japan	150	Venezuela	425

TABLE 5.3: Carbon emissions for steel produced around the world (per ton of steel, as a percentage of U.S. production).

10 kilos to 100 kilos
(20 pounds to 200 pounds)

A pair of shoes

1.5 kg (3.3 lbs.) CO_2e Crocs
8 kg (17.7 lbs.) CO_2e synthetic
11.5 kg (25.4 lbs.) CO_2e average
15 kg (33 lbs.) CO_2e all leather

> Imelda Marcos's collection of 2,700[1] pairs of shoes would have had a carbon footprint of around 30 tons, or 3 years' worth of 10-ton living—assuming, of course, that they had all been typical shoes.

As the numbers here show, shoes vary enormously in their carbon footprint (no pun intended). Just as important is their longevity.

At the low end of the carbon scale are Crocs, the simple and surprisingly durable shoe consisting of just 250 g of expanded EVA (ethylene-vinyl acetate) and sold without packaging. For these shoes, the raw material comes in at just over 1 kg (2.2 lbs.). The rest is a guesstimate.

The 8 kg (17.7 lb.) synthetic pair is based on a study of synthetic cross-country running shoes, made in China but traveling to market by boat. My average figure, meanwhile, is based on the input-output model (see page 191) and a price of $75 per pair. The model tells us that in the typical shoe about half of the carbon footprint is due to materials, around one-quarter is due to energy used in shoe manufacture, 15 percent is transport, 5 percent the shoe box, and 5 percent other bits and bobs.[2]

I have estimated the higher figure for all-leather shoes on the basis of the carbon intensity of cattle farming.

Most of our footwear comes from the Far East, although specialist leather might also have had to travel a long way to get there. Shipping is fairly efficient. The big inefficiency in transport comes if a product is air-freighted for speed. This is most likely in high-end fashion, though unfortunately there's no way to be sure as a consumer what has and hasn't been delivered from the country of origin by plane.

1 kg (2.2 lbs.) of cheese

12 kg (26 lbs.) CO_2e hard cheese

> That's about 3 kg (6.6 lbs.) CO_2e for a big 250 g (8 oz.) block from the store—equivalent to a 4-mile drive or a massive 12 kg (26 lbs.) of carrots.

It takes about 10 quarts of milk to make 1 kg (2.2 lbs.) of hard cheese, adding up to a considerable carbon footprint that's higher than that of many meats. The message is clear, then: going veggie doesn't reduce your impact if you simply swap cheese for meat. Neither will it save you money nor make you healthier. Perhaps the best advice if you're keen to reduce the climate impact of your diet is to think of cheese as a meat and therefore a treat. Many people will also improve their life expectancy by cutting back somewhat.

However much cheese you eat, there's an easy carbon win by keeping waste to a minimum. That means buying only what you think you'll actually get through and also avoiding trashing hard cheese just because it's showing a tiny sign of mold. This is perfectly safe according the U.S. Food Safety and Inspection Service, which must surely be among the most cautious groups around:

Discard any soft cheese showing mold. For hard cheese, such as Cheddar, cut off at least 1 inch around and below the mold spot (keep the knife out of the mold itself). After trimming off the mold, the remaining cheese should be safe to eat. Re-cover the cheese in fresh wrap and keep refrigerated.[3]

As for which hard cheese to buy, the most sustainable types probably come from cows that have grazed almost exclusively on rough pasture that couldn't have been used for crops—though of course that information isn't generally available in the stores.

Note that which country or area the cheese has come from doesn't matter much when set against the impact of the milk production (see page 72). Hence the easiest way to reduce the carbon footprint of your cheese is to opt for soft cheeses, because these require less milk to produce.

A congested commute by car

22 kg (49 lbs.) CO_2e five miles of crawling each way in an average car

> Every working day for a year would be 4.8 tons CO_2e more than flying from Los Angeles to Barcelona and back.
A congested drive can cause three times the emissions of the same drive on a clear road.[4]

Driving in a traffic jam very roughly doubles your fuel consumption per mile. However, that's only half of the story. By adding your car to the mass of ugly, belching motors, you also make a lot of other people line up just a little bit longer. It turns out, via a bit of simple queuing theory,[5] that the extra emissions you force everyone else to produce (when you add them all together) is about equal to the extra emissions that you produce yourself as a result of having to line up instead of being able to drive straight through. In other words, if your journey is congested, by choosing to do it you cause about three times more emissions than you might expect.

The queuing theory logic also works for the time that gets wasted. If you make the assumption that the journey is many times longer than it would be if there were no traffic, then the time you waste in the line is about equal to the sum of the extra time you make everyone else waste. In other words, the hassle and anguish that you experience is equal to the hassle and anguish that you inflict. So when deciding whether to drive through a busy area at rush hour, picture your own pain and double it.

All of this adds to the case for traveling by bike, bus, train, foot, or ride share wherever possible. It's also a useful reminder that all motorists should treat cyclists with the respect they deserve for helping to cut everybody else's journey time.

Where you must drive in busy conditions, do your best to minimize stops and starts—both your own and everyone else's. A steady slow stream of traffic is more efficient than a faster but less steady one unless the stops are so long that everyone can turn their engines off. One good tip is to think about what to do when two lanes merge: to reduce emissions, ease your speed down, merge gently and in good time, and allow others to do likewise. In theory at least, two lanes traveling at 50 mph can carry about the same traffic as 3 lanes traveling at 70 mph, assuming that everyone leaves a safe stopping distance between them and the next vehicle. This is because slower cars need less distance between them.[6] It's good to minimize the use of brakes on the highway if you can. And when you pass other vehicles, put your signal light on in good time too, so that no one else has to brake either.

A night in a hotel

3 kg (6.6 lbs.) CO_2e low-carbon scenario
24 kg (53 lbs.) CO_2e $100 spent on dinner, drinks, bed, and breakfast in a hotel with average eco-credentials
60 kg (132 lbs.) CO_2e high-carbon scenario

For my high-carbon scenario I have chosen one of those hotels where the TV and six lights are already on when you walk into your room. The room itself is too hot and you cool it by opening the window even though the radiator is on. There is a swimming pool, with air-conditioning. You order beef or lamb for dinner and it arrives with baby vegetables air-freighted from Peru. There is too much for you to eat. For dessert you have strawberries even though it is winter. In the kitchens, half of the food cooked is thrown out at the end of the night. You stay one night, finding your way through three towels as well as your sheets. You have a fried breakfast, giving the paper you ordered a quick glance before leaving it on the table (from where, surely, even in this hotel, it goes for recycling).

The low end of the scale could be a large, very well run hotel or, more likely, a simple bed and breakfast with thoughtful owners. If you stay a few nights, your sheets and towel aren't changed unless you ask. The room is comfortable, and you can adjust your own heating. You opt for a low-meat-and-dairy meal with seasonal vegetables, and you get to choose how much goes onto your plate. Leftovers end up in the next day's soup. You have something like cereal or muesli, fresh fruit and toast for breakfast. There is a selection of papers shared between guests (with the added advantage that you get to browse several if you have time). What you are paying for is a more personalized service in which you can have what you require without its being thrown at you just in case. The difference in carbon footprint between these two scenarios might be as high as a factor of 20.

The British clock up 42 million tons of emissions through their use of hotels, pubs, cafés, and restaurants (Figure 6.1). That's nearly 5 percent of the national carbon footprint. What the British drink when they are out has almost as much impact as what they eat, and both of these have a bigger footprint than the energy used by the establishments where the eating and drinking happens.

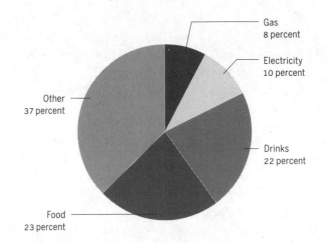

FIGURE 6.1: The 42-million-ton footprint of the U.K.'s hotels, pubs, and catering industry.

As a rule of thumb, the hotels, pubs, and catering industry in the U.K. has a footprint of about 400 g CO_2e for every pound you spend. North America is probably similar at around 270 g per dollar. Roughly speaking, this seems to be true whether it is food, drink, or accommodation that you are buying. However, this is just a general figure, and the footprint certainly doesn't have to go up or down with the price. Indeed, there is a lot that the carbon-conscious consumer can do to keep emissions down, simply by spending money in establishments that think about the issues.

When eating out, look for seasonal fruit and vegetables, and choose places where the lower-meat and lower-dairy options are cooked with at least as much passion as anything else. The restaurant should be taking steps to minimize food waste both on your plate and behind the scenes. In a hotel, look for good energy management, minimization of laundry and a general sense of care with resources. In a pub, look for local cask beer.

For any hotels, pubs, or restaurants seeking to understand their carbon footprint, a colleague and I have built and tested a carbon calculator especially for tourism businesses and have made it freely available online.[7]

A leg of lamb

38 kg (84 lbs.) CO_2e a 2 kg joint at the checkout
> For the same carbon footprint, you could have a bowl of porridge (made with half milk, half water) every day for 4 months.

Lamb comes in with a carbon footprint of about 17 kg for each kilo (7.7 kg for each pound) produced at the slaughterhouse. Transport, basic processing, refrigeration, and a little bit of packaging each add a little bit, so that by the time the meat reaches the checkout, the footprint has increased by about 10 percent. You will add a similar amount again by the time you have picked it up from the store, put it in the fridge and cooked it, taking the overall carbon impact to more than 20 kg per kilo (9 kg CO_2e per pound).

The issues surrounding sheep are very similar to those relating to cows (see Steak, page 97, and Milk, page 72). Like cows, sheep ruminate, releasing large quantities of methane. And just as with beef farming, the exact impact of different types of sheep farming is complex and only partly understood. Hill farmers can claim that they are putting otherwise unproductive land to use. Some also claim that putting sheep on the hills helps the soil to absorb carbon from the air. Counterarguments are that hill-farmed sheep are inefficient, that they spend too much energy wandering around, eating low-energy food and keeping warm and that therefore they burp more methane per joint of meat than their lowland counterparts.

It seems probable that, from a broad sustainability point of view, hills are the best places to have sheep. But ultimately only one thing is clear: a low-carbon world is going to have to involve less lamb. The typical footprint of this meat is even higher than that of beef. The low-carbon choice is to think of lamb as a treat and to eat less of it.

A carpet

76 kg (168 lbs.) CO_2e thin polyurethane carpet with thin
underlay, 4 m × 4 m (13 ft. × 13 ft.)
290 kg (639 lbs.) CO_2e the same area covered in thick wool,
polypropylene, or nylon with generous underlay

> If you have 50 m² (540 sq. ft.) of top-end carpet in your house,
> that could add up to 900 kg of embodied emissions—equivalent
> to a burger a day for a year.

Provided you get full wear out of them, some carpets may
well pay for themselves in carbon terms, by improving your
insulation. However, if you are in the habit of moving house
every couple of years and insisting on stripping out everything that was there to replace it with styles more to your
own taste, then carpets, along with other soft furnishings,
could be a significant chunk of your total carbon footprint.

Table 6.1 gives some figures for the footprints per pound
of fabric materials, based on studies of European production.[8] In the U.K. most textiles come from developing countries, not least China, where industry is typically a lot more
carbon intensive. For Chinese production you can probably
mark most of these up by a factor of two or three on the basis

Carpet type	Carbon footprint (kg CO_2e per kilo)	(kg CO_2e per pound)
General	3.89	1.77
Felt underlay	0.96	0.44
Nylon	5.43	2.47
PET (polyethylene terephthalate)	5.55	2.5
Polypropylene	5.03	2.29
Polyurethane	3.76	1.71
Wool	5.48	2.49

TABLE 6.1: Carbon footprints of carpet types.[9]

that the factories tend to be less energy efficient, and the electricity they use is also more carbon intensive per unit because so much of it comes from coal-fired power stations. I'm not writing this out of some kind of protectionist instinct—just presenting the facts as I see them. I'd like to see China develop—but not at any cost.

To give a sense of what the numbers mean in practice, typical weights are 1 to 1.5 kg per square meter (93 to 140 g per square foot) for underlay and 1 to 3 kg per square meter (93 to 280 g per square foot) for the carpet itself. This puts the overall footprint in the region of 5 to 18 kg per square meter (1 to 3.6 lbs. per square foot).

Using a cellular phone

47 kg (103 lbs.) CO_2e a year's typical usage of just under 2 minutes per day

1,250 kg (2,760 lbs.) CO_2e a year's usage at 1 hour per day

125 million tons CO_2e global cell phone usage per year

> A minute's cell-to-cell phone chatter comes in at 57 g,[10] about the same as an apple, most of a banana, or a very large gulp of beer. Three minutes have a similar impact to sending a small letter (written on recycled paper) by regular post.

Cell phones cause a fairly tiny slice of global emissions, but if you are a chatterbox using your cell phone for an hour each day, the total adds up to more than 1 ton CO_2e per year—the equivalent of flying from London to New York, one way (economy class).

Indeed, the footprint of your cell phone use is overwhelmingly determined by the simple question of how often you use it. One estimate for the emissions caused by manufacturing the phone itself is just 16 kg (35.3 lbs) CO_2e,[11] equivalent to nearly 1 kg (2.2 lbs.) of beef. If you include the power it consumes over two typical years (that's about how long the average phone remains in use, even though most could probably

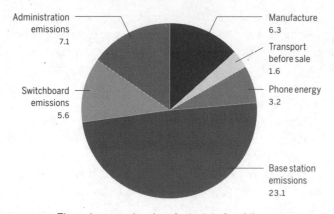

FIGURE 6.2: The 47 kg annual carbon footprint of mobile usage, based on a Nokia N7600 phone used for 2 minutes per day and replaced after 2 years.[13]

last for 10 years) that figure rises to 22 kg (49 lbs.).[12] But the footprint of the energy required to transmit your calls across the network is about three times all of this put together, taking us to a best estimate of 94 kg (207 lbs.) CO_2e over the life of the phone, or 47 kg (103 lbs.) per year (Figure 6.2).

In 2009 there were 2.7 billion cell phones in use: nearly half the world population has got one. On this basis, cell phone calls account for about 125 million tons CO_2e, which is just over one-quarter of a percent of global emissions.[14]

If you want to reduce the footprint of your communication habits, texting is a much lower-carbon option (see page 16). Landlines offer carbon savings, too, because it takes about one-third of the power to transmit a call over a fixed landline network than it does when both callers are on a cell phone.[15]

It took a lot of digging to get data for this entry. In the end I was pleasantly surprised that there is some reasonably sensible-looking analysis out there. Nevertheless, now feels like a good time for another reminder that all footprint estimates contain considerable uncertainty, and some even more so than others.

Being cremated

80 kg (176 lbs.) CO₂e

> That's less than one ten-thousandth of your life's carbon footprint.

My advice would be to treat yourself on this one occasion to whichever method takes your fancy. This is the one time when it is too late to start worrying about your carbon footprint. And anyway you have already done the most carbon-friendly thing possible. However, this book is about doing the numbers, so here goes.

The *Guardian* reported that the average cremation uses 285 kilowatt-hours of gas and 15 kilowatt-hours of electricity, and I'm going with that. I have not included emissions from your own flesh, because your body was only ever a temporary carbon storage device.* On top of the carbon, cremation sends significant amounts of mercury into the atmosphere.

Burial sounds like a more climate-friendly solution, but browsing blogs on this subject (yes, there really are people who write these) I found a clergyman who reckons that burial turns out 10 percent higher carbon once you take account of cemetery maintenance for the next 50 years. I have managed to resist checking these sums myself.

A sea burial sounds ideal if you don't mind the possibility that some of your loved ones may be heaving over the side when they are supposed to be paying you their last respects. There are usually legal problems with this method, however.

If what you most want to do is send a final eco-message to the world, the best answer I know of is to be dressed in easy-to-rot clothing and put in a wicker coffin. In some countries it is possible to be buried in woodland with the idea that your remains will become trees—a lovely idea, though if everyone tried this we might run out of room. Check the legality In your part of the world if this is of interest.

---

* If we start getting info this, losing weight will become a source of guilt. Perhaps even a bit of reverse psychology will kick in and alleviate the western obesity pandemic.

100 kilos (220 pounds) to 1 ton

New York City to Niagara Falls (405 miles) and back

53 kg (117 lbs.) CO_2e banana-powered bike
66 kg (146 lbs.) CO_2e coach
120 kg (265 lbs.) CO_2e train
330 kg (728 lbs.) CO_2e small, efficient car
500 kg (1,100 lbs.) CO_2e plane
1,100 kg (2,530 lbs.) CO_2e large four-wheel drive

All these scenarios are based on one person traveling each way on his or her own. I've based the figures for the small efficient car on a Fiat 500 traveling at a steady 70 miles per hour and getting 45 miles per gallon. The four-wheel drive, meanwhile, is based on a Land Rover Discovery doing 16 miles to the gallon. If it goes above 70 miles per hour or puts the air-conditioning on, its impact will be higher still.

For all the road vehicles, the exhaust-pipe emissions make up about half of the footprint. About one-third lies in the manufacture and maintenance of the vehicle itself, and the remaining one-sixth is down to the supply chain of the fuel (see Gasoline, page 89). I've assumed that you keep to the speed limit and look after your car with about average care.

The bike is the outright winner if you can afford the time, you are careful about what you eat (see Cycling, page 26), and you don't have a headwind. Of the more practical options, the coach comes out on top, with a footprint more than 15 times smaller than the gas guzzler. One reason that the coach beats the train is that it travels more slowly, which is significant because the energy needed to overcome air resistance goes with the square of the speed. Another reason is that although a coach is heavy, the weight per passenger is much less than it is for a train (see page 42).

Some analyses that I've seen put a train ticket and a solo drive closer together in carbon terms. But I'm suspicious of these claims because the embodied emissions of the car per passenger mile are often ignored or underestimated. Whatever the precise difference (and it will of course vary widely depending on the particular vehicles), the train also lets you get some work done, read a book, or sleep instead of arriving at the other end stressed and frazzled.

The plane could actually be better than driving if you have the wrong kind of car. (My sums are based on flying economy class.) But please don't take this as an advertisement for flying: it's just a reminder of quite how carbon-profligate some road vehicles are.

As soon as there are more people on the trip, of course, cars become a lot more efficient. If we load the whole family into my c_1, along with everything for a week's holiday and put bikes on the back (it is possible, but only just), the fuel consumption goes down by at least 10 percent. But the emissions per passenger fall so low that we'd be better going that way—in carbon terms, at least—than all traveling by train.

When it comes to both speed and safety, trains and planes win. When you are calculating how much of your life will be taken up by the journey, my back-of-the-envelope calculations tell me that a driver with a fairly typical life expectancy should add about 2 hours each way to the car journey time to take account of the 1 in 200,000 chance that they will lose the rest of their life in a crash.[1] If you are in your twenties and in good health you might want to call it 3 hours. This is a very significant chunk to add on to the expected journey time of 7 hours[2]. For trains and planes the average loss of life expectancy through injury or death is vanishingly small, despite the lavish media coverage that any crash does get. I'm sad to have to report, for the sake of even-handedness, that the bike will lose hands down on safety grounds unless you are careful with your route choice.

A common myth is that huge four-wheel-drive guzzlers are safer for their occupants. This is generally not true. They are, however, more dangerous for everyone else on the road.

Christmas excess

4 kg (8.8 lbs.) CO_2e per adult low-carbon scenario
280 kg (617 lbs.) CO_2e per adult U.K. average
1,500 kg (3,300 lbs.) CO_2e per adult high-carbon scenario
> A full-on Christmas could cost you a couple of months' worth of 10-ton living.

I said at the beginning that this book was about picking your battles. Christmas has got to be a good place to go looking, even if it might entail breaking a few habits and engaging in some delicate family negotiations. For most of us there is a golden opportunity here to escape some mindless consumerism, stress, and perhaps even debt.

In my numbers I have only included unwanted presents, wasted food, avoidable travel, Christmas lights, and cards. Clearly it's not a complete list, but it's enough to give a flavor. The numbers are per adult and are based on three scenarios, none of which is intended to be ridiculous.

The average U.K. adult spends a massive £440 ($660) on presents, of which 20 percent will be totally unwanted.[3] The U.S. figure is similar at $430 per person (not just adults).[4] There will also be a lot of "partly wanted" middle ground, so I've assumed an average "wantedness factor" of 50 percent for all presents. In the festive season we spend about £150 ($225) more than usual on food, and I've allowed one-third

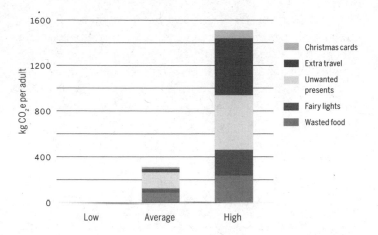

FIGURE 7.1: The footprint of Christmas waste in the three scenarios.

for waste, thinking that this will be slightly higher than it is in the rest of the year because of the "Oh-no-not-turkey-again" effect and the fact that the big meals tend to keep coming over the whole period long after most of us have reached our "wafer-thin mint" threshold.[5] The Christmas lights burn through about 45 kilowatt-hours. The average adult mails about 20 cards, with most of the footprint coming from the delivery, not the paper. We typically travel 50 miles each above what we would do anyway, and it is generally by car.

In the high-carbon scenario, you spend $1,500 on presents (yes, that feels extreme to me, too, but it's only a little over double the average). Sadly, in this scenario the "want-edness factor" turns out to be just 30 percent because you are even worse than me at choosing presents. People are too embarrassed to tell you or to sell them, so they gather dust or even get sneaked into landfill. You decorate your house with a wild lighting display that doesn't use LED bulbs. You mail 200 rather large cards. You also clock up 500 miles on a tour of relatives in a thirsty car.

I think the low-carbon scenario could be at least as festive and a lot less hassle. The food is great but none gets wasted.

You might eat a bit too much, but you make up for that over the coming months, so it's not additional. Your presents are thoughtful but not necessarily expensive. You encourage people to be honest in their reaction and you've kept all the receipts. You have LED Christmas lights. You stay at home and you send cards only to a few people that you haven't seen for ages and with whom you really don't want to lose touch. You video-Skype your distant relatives and make plans to see them properly another time.

Some British friends of ours spread the word that only children were going to get presents worth more than a strict limit of £1 ($1.50) . They asked everyone to reciprocate, packing any cash saved off to the charity of their choice. Both giving and receiving became an exchange of gestures and altogether more fun.

Insulating an attic

350 kg (770 lbs.) CO_2e outlay for a three-bedroom house
35 tons CO_2e payback over 40 years
> The payback of insulating an attic can be a remarkable three and a half years' worth of 10-ton living.

My calculations are based mainly on figures produced by the U.K.'s Energy Saving Trust[6] and assume you are adding 270 mm (10 inches) of insulation to the previously uninsulated attic of a three-bedroom house. According to the EST's figures you save 800 kg (1,760 lbs.) CO_2e per year, but I've rounded this up to 880 kg (1,940 lbs.) to take account of fuel supply chains that I know they don't include.

The embodied energy of the insulation material pays back in less than six months and is good for at least 40 years. You will therefore save about 35 tons of greenhouse gas.

In terms of money, even without a government grant, you'll get payback on your $750 investment in 4 years, even when a 10 percent discount rate is applied. In other words, the decision to insulate your attic tomorrow will save you

$1,450 on top of paying back your outlay compared with investing the money in a bank account with a 10 percent interest rate. (See page 131 for more on discount rates.) In other words, it's a no-brainer. In the U.K., the EST may well offer you a 50 percent grant, too, which makes it a no-brainer even if you are suspicious that they may have been optimistic with their numbers.

Table 7.1 gives a detailed breakdown for the scenario discussed so far and also for someone increasing their insulation from 50 mm to 270 mm (2 to 10 inches). This is a good move, too, but only if you care about the carbon savings or can get a grant. If you are just in it for the money, and you apply a discount rate, then I don't think you ever quite get it back again. However, at just $7.5 per ton, the CO_2e saved improving your existing insulation is still a hugely cost-effective way of investing in a lower-carbon world.

The EST's calculations that I've used here are based on the assumption that rather than cashing in on all the financial and carbon savings that would be possible if you kept your home at the same temperature that it used to be, you will in fact allow your home to be warmer once it is insulated. In

	From no insulation to 270 mm (10 inches)	From 50 mm (2 inches) insulation to 270 mm (10 inches)
Cost without a grant	$775	$775
Annual payback	$225	$225
Embodied emissions in the material[7]	380 kg/830 lbs.	380 kg/830 lbs.
Annual carbon savings (including fuel supply chains)	880 kg/1,940 lbs.	880 kg/1,940 lbs.
Financial payback period (with 10 percent discount rate applied)	4 years	Never quite makes it
Payback period (with 10 percent discount rate applied)	$1,400	–$75
40-year carbon savings	35 tons	10 tons

TABLE 7.1: Insulating the loft in a three-bedroom house without a government grant: the money and the carbon.

other words the sums here assume that you will lose some of the available savings in exchange for a warmer and perhaps more comfortable home.

Various types of attic insulation are available: you can get the standard synthetic kinds as well as varieties from sheep's wool, paper, and a range of other options. Some of these sound good, but you should choose them only if you are 100 percent convinced that there is no compromise on performance or the longevity. Those are the priorities.

A necklace

Zero CO_2e handed down or made from driftwood and seashells
200 kg (440 lbs.) CO_2e $750 worth of new Welsh gold
400 kg (880 lbs.) CO_2e $750 worth of gold and diamonds
 sweated out of mines in developing countries

Who would have thought that something so small could have such an impact! But think about it for a moment, and it makes sense: gold and diamonds are precious precisely because it takes effort and sweat to extract them.

At the bottom of my scale are items for which the value is in the art and not the materials. Also at the low end of our scale is a piece of jewelry that has been passed on or reforged from an existing item. The carbon impact here is simply from the energy required to melt it down.[8]

To arrive at my ballpark figure for the carbon footprint of jewelry—265 g CO_2e per dollar spent—I have once again used the technique of working out the carbon footprint of an industry and dividing it by that industry's total output. The same model that we used to get the overall figure can give us an idea of where that footprint comes from. Not surprisingly, it turns out to be attributable to the extraction of metals and minerals, such as gold and diamonds.

For my "average" example, I have chosen a necklace from virgin Welsh gold, simply because although it has been mined especially for you, this has been done using relatively efficient mining technologies and in a country where machinery tends to be more fuel efficient than in developing countries. The price of the Welsh gold will reflect the relatively high fuel taxes in the U.K., and this reduces the footprint per dollar spent somewhat.

At the top end of the scale is jewelry obtained using inefficient technology and cranky machinery. My figure of 530 kg (1,168 lbs.) CO_2e per $1,000 is simply a crude estimate based on twice the carbon intensity of typical U.K. industry.[9]

While on the subject of gold and diamonds, I should mention that chunks of the Amazon are being deforested in the pursuit of gold. Poor people in developing countries are being exploited in the production of both gold and diamonds. Is it worth it? Can it really be a romantic gesture to give someone something that has an embodied footprint of exploitation? Or can there be beauty and elegance without the extravagance?

A computer (and using it)

The machine itself
200 kg (440 lbs.) CO_2e a simple low-cost laptop
720 kg (1,590 lbs.) CO_2e a 2010 21.5-inch iMac
800 kg (1,760 lbs.) CO_2e an all-the-frills desktop

Electricity consumption
13 g CO_2e per hour an energy-efficient laptop[10]
69 g CO_2e per hour a 2010 21.5-inch iMac
165 g CO_2e per hour an old desktop machine

Your use of servers and networks
Typically 55 g CO_2e per hour
> This is the fastest-rising part of the footprint of computing (see Data centers, page 160). Add a bit more for any peripherals and the demands you place on other machines via your use of the Internet.

The machine itself

Even before you turn it on, a new iMac has the same footprint as flying from Glasgow to Madrid and back.

Apple has carried out a detailed life-cycle carbon assessment of their business and their products.[11] This analysis suggests that the company's mid-sized desktop machine—the 21.5-inch iMac—comes in at around 570 kg (1,250 lbs.) CO_2e.[12] However, the devil is in the details, and the life-cycle approach that Apple has used has a nasty habit of "leaking"—missing little bits of the footprint. The footprint of a computer comes from the complex mass of activity that has had to go on throughout the economy in order for minerals in the ground to turn into machines in the stores. Each component is in turn made of materials and other components, behind each of which lies a whole life cycle of its own. To trace this by mapping out the different processes one by one is impossible because the ripple effect is mathematically endless. You have to miss some processes out, cutting the pathways short, and the result is a shortfall in your footprint calculations that is known as "truncation error." And it's the reason that I think Apple's figure is almost certain to be a little on the light side.

The "input-output" approach of tracing carbon impacts through the economy by modeling the way in which industries buy and sell from each other has, for all its generalizations, the huge advantage of not *systematically* missing bits out. Based on a 21.5-inch iMac costing $1,800 in the store, input-output modeling gives me a footprint estimate of about 720 kg (1,590 lbs.) CO_2e. Just as expected, that is a bit higher than the figure produced by Apple's process-based approach, so it is the one I have gone with.[13]

Apple, on its website, talks about reducing its impact by making machines lighter, but the bulk raw materials are just a small part of the issue. If a laptop were just a lump of plastic, steel, and semiconductor, you could get its footprint to below 50 kg (110 lbs.) CO_2e. The problem is that microprocessors come in at around 5 kg (11 lbs.) CO_2e for a 2 g chip because of the high-tech process that is involved and the incredibly clean environment that is needed.[14] Apple also talks about reducing packaging; this, too, is good practice but makes a marginal difference in the scheme of things.

It's hard to give guidance on what makes a low-carbon computer because the processes involved in making one are so complex. The guidance we would get from input–output analysis is that the cheaper your machine, the less its footprint is likely to be. This is probably a reasonable rule of thumb, although it may mask the impact of some cheap, carbon-intensive production in developing countries. Another guiding principle could be to choose the products from a country that has efficient industry and not too much reliance on coal for electricity—but this is tricky because assembly might take place in a country other than the one in which the most energy-intensive 2 gram components are made.

Using the computer

The electricity emissions typically equal the footprint of manufacture after 15,000 hours—that's 9 hours every day for 5 years.

Apple reports that the iMac we're talking about consumes 91 watts of electricity in use.[15] The company also reports that the power supply is 87 percent efficient, so, if I understand them correctly, that makes a total of 105 watts leaving your plug. If that is right, in the U.K. the emissions from use would equal my estimate of embodied emissions after 11,500 hours (that's 7 years at 8 hours per day for 200 days of the year), and by this time the cost of the electricity will probably also have worked out about the same as you paid for the machine. In Australia it would take just 7,000 hours because more of the electricity comes from coal (see page 193). In the U.S. it would take 10,500 hours and in China about 8,000. Most people would change their machine before clocking up those hours, so the embodied emissions in the machine are the biggest deal.

However, the sums don't necessarily always work out like that. The iMac is a high-value computer, and I have associated that with a relatively high footprint. In addition, some machines are a lot more power hungry. Apple says they have worked on becoming more efficient. Traditionally, laptops consume less than desktops, because it has always been important to conserve battery life. Some, but not all, desktops are catching up. I recently encountered an office full of fairly new HP and Compaq PCs that were burning through 24 watts even when they were switched off. Since they were only on during office hours, they were consuming more when turned off than on. The answer was simply to unplug them at the end of the day.

I haven't taken into account the use of peripherals or the activity you might stimulate in other machines around the web through your emails and web searches (see Data centers, page 160). Despite all this, computing can be a fairly low-carbon way of spending time.

To summarize, computing could be a few percent of your carbon footprint. The embodied footprint of a computer is significant and could easily be the dominant factor, so it probably doesn't make sense to buy a new, more efficient machine on carbon grounds—better to make an old one last. However, when choosing a computer, do think about its power consumption, especially if you will use it a lot. Laptops are still usually better in this respect than desktops, but whatever you use, switch it off when not in use and unplug it if that's what it takes to get the power to zero.

A mortgage

800 kg (1,760 lbs.) CO_2e per year for $150,000
on 5 percent interest
> That's a whole month of 10-ton living.

How can a mortgage have a carbon footprint? Surely it just boils down to a few bits of paper or electronic transactions?

Look more closely. The bank or building society runs offices, buys computers, sends mail (probably mainly junk; see page 45), and stores data. Its people travel. It outsources everything from cleaners to building maintenance, from design work to corporate lunches, and maybe even still buys in the odd paper clip.

What I am saying is that when you take out the loan, you feed the financial services industry along with all its direct and indirect environmental impacts. This is another example of a set of ripple effects across the economy that we can't see and don't stand a chance of counting up one by one. Happily, our input–output model (see page 191) comes to the rescue and produces a ballpark figure of 106 g CO_2e for every dollar spent on financial services.

If you have a $150,000 mortgage on a 5 percent interest rate, you pay $7,500 per year (plus any actual repayments) and this incurs an annual footprint of the order of 800 kg (1,760 lbs.) CO_2e. The same story applies to all loans, and the principle goes wider still. All the intangible services have fairly similar carbon intensities: solicitors, lawyers, accountants, therapists, architects, and so on.

There are two basic lines of attack if you want to cut the carbon. The first is to take out a smaller mortgage and spend the money you saved on something that decreases carbon emissions, such as an investment in an offshore wind farm, a "save the rainforests" project, or, if you want your neighbors to know what a good person you are, a solar roof. You could stick the money in the bank where it may seem harmless, but even then you may be enabling the bank to lend more to profligate consumers. The other line of attack is to be discerning about the way the mortgage company goes about its affairs. I have based the footprint estimate on general figures for the industry, but actually there will be good and bad practices within it. To begin with, one-tenth of the sector's footprint comes from printing and postage, so supporting a bank or building society that doesn't do junk mail is a good first step.

About 30 percent of the industry's footprint comes from air transport, but I'd be surprised if, for example, the Ecology Building Society, based in Yorkshire, goes in for much of this. They run a simple, lean operation out of eco-friendly premises and make a real effort to walk the walk. If I had to guess, I'd put their carbon intensity at less than half of the industry average. Furthermore, their footprint is in the cause of encouraging a sustainable building stock, because they vet their loans by the sustainability of the project and also support lenders in improving their buildings.

The job of choosing between more mainstream lenders is trickier. The most important question is probably to ask yourself how much you *trust* them to take the carbon issue seriously. If the answer is that you don't, then they probably haven't done much to be any better than the industry norm, no matter how much they are talking about it. That is my experience in most of the industries I work in.

1 ton to 10 tons

A heart bypass operation

1.1 tons CO₂e

> That's nearly six weeks of a 10-ton lifestyle—equivalent
to a couple of return flights from San Francisco to Vancouver.

The carbon cost of health care in the U.K. is around 170 g per
pound sterling (110 g per US$1) spent. Although the U.S. and
Canada have very different health care systems, the carbon
intensities are probably reasonably similar. If so, looking after
your health turns out to be a fairly low-carbon way of spending money. And in terms of the quality of life improvements
we stand to gain from it, health care when we need it must be
one of best ways of spending our carbon budget.

That said, a big operation clocks up a big footprint. The
typical cost of a heart bypass to the U.K.'s National Health
Service is £6,324 ($9,500).[1] If we assume that this operation
is averagely carbon intensive U.K. health care spending, that
adds up to more than 1 ton CO_2e.

Overall, U.K. health care has a footprint of 27 million tons
CO_2e, or just over 3 percent of the national consumption footprint. Electricity and fuel used by health services account for
less than one-third. Drugs account for nearly one-fifth. My
catch-all "other" category is nearly one-third of the total,
reflecting the variety of equipment and other stuff that is

required to keep us healthy. Paper and cardboard surprised me at a massive 2 percent of the footprint of all health care. I'd like to think this is not the stuff that clogs up the filing cabinets of one of the world's biggest bureaucracies but rather the consumables used to keep things clean. So what can we do to reduce the emissions of our health care? The best option is to stay healthy, of course. This might involve cycling (safely) or walking more, and thinking about the amount of meat and dairy in your diet—all things that will reduce your direct footprint, too, and are discussed elsewhere in this book. When you do actually need health care, be careful with medical resources, but relax in the knowledge that at around 110 g CO_2e per US$1, it is one of the lower-carbon ways for you or your government to spend money.

Photovoltaic panels

3.5 tons CO_2e producing a solar roof capable of generating
1,800 units (kilowatt-hours) of electricity per year
50 tons CO_2e lifetime savings—that's 5 years' worth of
10-ton living
> WARNING: This section contains myth-busting payback
calculations that will interest some more than others.

I'm going to do the financial sums and the carbon sums and then put these both together to see how electricity-producing photovoltaic solar panels rate as a cost-effective way of saving carbon.

First, the financial bit. Some governments offer a "feed-in tariff" to reward individuals who install solar panels on their roofs. In the U.K., for example, homeowners are offered a massive 36.5p (57 cents) per unit generated.[2] This handout is guaranteed for the next 20 years. On top of the feed-in tariff you can still use what you generate yourself (thus cutting the amount you have to buy) or sell it back to the grid to get even more revenue. It's an incredibly generous government handout (especially given the U.K.'s financial situation), and if currently available micro-photovoltaic panels are a viable source of electricity, surely we should all be diving in?

U.K. analyst and author Chris Goodall[3] has done sums on the financial payback from micro-renewables. He estimates that in the U.K. it will cost you £10,000 ($15,000) to get a set of panels installed that is capable of providing you with 18,000 kilowatt-hours per year. Once you have taken account of income from the tariff, your sales to the grid and reductions in your grid electricity bill as well as annual maintenance costs, Chris thinks you can make a return of £730 ($1,125) per year. This figure suggests a financial breakeven after 14 years. That sounds fine, but what this is really saying is that provided everything goes to plan, you will be exactly as well off as you would have been if you had kept the money *in a box under your mattress.* Such a simple "payback period" calculation would be fatally flawed because it would ignore both the fact that you could have done something else with the money, where at the very least you would have gotten a bit of interest to offset inflation, and the fact that even the surest-looking projects, backed up by manufacturers' guarantees, carry a degree of risk.

More realistic payback sums need to have a way of taking into account the fact that money in your hand right now is worth more to you than the promise of the same amount of money to be paid to you in the future provided that things go well. This can be done by applying a so-called discount rate to the future payback. Applying a 10 percent discount rate (a fairly sensible figure) is equivalent to saying that you'd be just as happy to have $900 in your pocket now as you would be to have $1,000 promised to you in a year's time on the condition that your photovoltaic panel project is still going according to plan. If we following the same logic, a promise of $1,000 in two years' time is worth just $810 to you today, and the financial return that you hope to get in your 14th year is worth less than one-fifth of the same money in your hand right now. So, what happens to your solar payback period once a 10 percent discount rate has been applied? It turns out that you would never get more than two-thirds of your money back, even if your panels lasted forever. (Which they won't. After 20 years they can be expected to be functioning at less than 80 percent efficiency, and after 40 years they will probably have had it.) In other words, don't buy a solar roof purely as a profit-making venture, even with the government's wildly generous feed-in tariff.

If your government or state only pays you, at most, the market rate for the electricity you feed into the grid (as is the case throughout North America as I write), the finances are likely to be disastrous.

But what about the carbon sums? I'll guesstimate that the £10,000 ($15,000) you spend is half on the kit and half on the installation. To give the carbon sums their very best possible chance, I'll generously overlook the footprint of installation and use the lowest plausible figure I can take from my input–output model for the manufacture of the panels: 0.47 kg CO_2e per dollar spent. That gives the panels a footprint of 3.5 tons. If we assume that the electricity generated all replaces output from coal-fired power stations rather than the grid average, then the carbon savings per year is about 1.8 tons, and you'd pay back the carbon in about 2 years. So where does that leave us? After 40 years your net cost (your initial investment minus the paybacks each year with discount rate applied) is still over £3,000 ($4,600). The government will have invested £13,000 ($20,000) over the 20 years of the feed-in tariff and (I'm assuming) nothing from then on. Something like 50 tons CO_2e will have been saved.[4] That's a cost of £330 ($500) per ton, even worse than a micro wind turbine and dramatically worse than offshore wind.

Are there any reasons to get a solar photovoltaic roof? Perhaps. You might want to invest in a developing technology. Or you might simply want one for fun. If you need to buy things to prove your status in society, solar panels are one of your most carbon-friendly options. We spend billions on mindless junk and flights around the world for that very reason: status. With the panels you can show everyone that you have spare cash but that you also think about the world. Photovoltaic panels can replace the SUV, and you might still be in the vanguard of this trend if you are very quick.

Flying from Los Angeles to Barcelona return

3.4 tons CO_2e economy class
4.6 tons CO_2e average
13.5 tons CO_2e first class[5]

> Three economy trips are a whole year's worth of 10-ton living.
One trip is equivalent to 340,000 disposable plastic carrier bags.

In other words, for your plastic bags to have the same footprint as just one trip from L.A. to Spain, you would have to go to the supermarket every single day for 10 years and return each time with 93 disposable bags. A flight from New York to London has roughly half the impact. The distance is a bit more than half, but there is a slight efficiency gain because there is less fuel to carry. New York to Vancouver or San Francisco is just over a third of the distance.

A Boeing 747 carrying 416 passengers burns through 116 tons of fuel on the 9,700 km (6,030-mile) flight each way. Almost one-third of the total weight on take-off is fuel. As the fuel burns, it creates three times its weight in CO_2. But the impact is worse still because high-altitude emissions are known to have a considerably greater impact than their low-altitude equivalents. The science of this is hideously complex and poorly understood,[6] but there is still a clear case for applying a multiplier to aviation emissions to take account of their extra impact. I have used a factor of 1.9.[7] Aviation is sometimes said to account for between 1 and 2 percent of global emissions. These statistics ignore the effect of altitude. The proportion is also higher in the developed world, especially in those bits of it, like the U.K., that are surrounded by sea. British travelers' personal flights account for a huge 8 percent of the carbon footprint of all consumption. That rises to nearly 12 percent once business flights and air-freight are added on.

In terms of your own lifestyle it might be much less than this. Many people never fly at all. Then again, for some people, flying accounts for the overwhelming majority of their total footprint, and trying to cut carbon in other areas might simply be a misdirection of attention, distracting them from what matters. First-class and business-class tickets are particularly high in impact simply because your seat uses up more of the plane and because by paying more money you provide a greater proportion of the commercial incentive for the flight. It's hard to imagine a low-carbon flying technology coming to the rescue. The physics of flight simply does not allow us to reduce the energy it takes to keep us in the air by more than a few percent,[8] and for the foreseeable future that energy has to come from fossil fuels. Nevertheless, there are still some efficiencies to be had. One of these is the automation of air-traffic control to replace the current archaic manual system. Humans are woefully unable to calculate optimum flight paths in real time with hundreds of planes in the air at once, all competing for space and time slots. One estimate is that upwards of 9 percent efficiency improvements are possible.[9]

Ultimately, then, it's hard to avoid the conclusion that most of us need to fly less. But that needn't make our lives any worse. Make your flights count: go for longer but less often, and do things you really couldn't do at home. For the rest, try local trips, which involve less travel time and therefore more holiday. After all, the experiences of getting to an airport, hanging around in a departure lounge, and then sitting cooped up for hours are intrinsically terrible ways of spending time. Also think about *where* you fly to: the closer the destination, the fewer the emissions. One myth is that long-haul flights are automatically more efficient per mile than short-haul because they involve proportionally less time taxiing, lining up, taking off, and landing. This isn't necessarily true, because the long-haul flight has to lift more fuel. The truth is that the most carbon-efficient way of getting across the world is in several hops—but not too many.[10] But none of this changes the fact that the further you fly, the larger the footprint.

Of course, the flying conundrum affects companies as well as individuals. I work with a few businesses for whom flying is a key issue. They know it's high in carbon, costly, and time consuming. They also know they have always had strong business reasons for doing it. New thinking is required to break out of old habits. Video conferencing may never fully replace human contact, but it is a lot cheaper and easier once you are fully conversant with the technology. What is worth more, one face-to-face visit or ten video link-ups?

It is difficult to see a place in the low-carbon world for much air-freighted food (see Asparagus, page 85), let alone durable goods such as clothing. Some garments are air-freighted simply to reduce lead times and cut the cost of stock that is tied up in transit at sea. Air-freight labels are one piece of consumer information that would surely be simple and helpful. Currently these are found on some supermarket fresh produce but nowhere else.

I'm sometimes asked about air freight from developing countries: "Surely it's good to keep supporting that country by carrying on the trade!" In broad terms, I don't think so. The argument is a bit like saying that you should keep the arms trade booming so that people can keep their jobs. Economies need to be powered by people doing things that are useful. Anything else is an unsustainable nonsense.

1 ton of fertilizer

2.7 tons CO_2e nitrogen fertilizer efficiently made and
sparingly spread

12.3 tons CO_2e the same stuff made inefficiently and
used in excess

> A real carbon opportunity: up to half a percent reduction in
global emissions—it's dead easy and has no bad side effects.

Nitrogen fertilizer is a significant contributor to the world's
carbon footprint. Its production is energy intensive because
the chemical process involved requires both heat and pressure. Depending on the efficiency of the factory, making 1
ton of fertilizer creates between 1 and 4 tons CO_2e. When
the fertilizer is actually applied, between 1 and 5 percent of
the nitrogen it contains is released as nitrous oxide, which is
around 300 times more potent than CO_2. This adds between
1.7 and 8.3 tons CO_2e to the total footprint,[11] depending on
a variety of factors.[12] Here's how the science of it goes. All
plants contain nitrogen, so if you're growing a crop, it has
to be replaced into the soil somehow or it will eventually
run out. Nitrogen fertilizer is one way of doing this. Manure
is another. Up to a point there can be big benefits. For some
crops in some situations, the amount of produce can even be
proportional to the amount of nitrogen that is used. However, there is a cut-off point after which applying more does
nothing at all to the yield, or even decreases it. Timing matters, too. It is inefficient to apply fertilizer before a seed has
had a chance to develop into a rapidly growing plant. Currently these messages are frequently not understood by small
farmers in rural China, especially, where fertilizer is as cheap
as chips and the farmers believe that the more they put on
the bigger and better the crop will be. Many have a visceral
understanding of the needs for high yields, having expe-
rienced hunger in their own lifetime, so it is easy to understand the instinct to spread a bit more fertilizer. After all,
China has 22 percent of the world's population to feed from

9 percent of the world's arable land. There are other countries in which the same issues apply, although typically the developed world is more careful. Meanwhile in parts of Africa there is a scarcity of nitrogen in the soil and there would be real benefits in applying a bit more fertilizer to increase the yield and get people properly fed. One-third of all nitrogen fertilizer is applied to fields in China—about 26 million tons per year. The Chinese government believes there is scope for a 30 to 60 percent reduction without any decrease in yields. In other words, emissions savings on the order of 100 million tons are possible just by cutting out stuff that does nothing whatsoever to help the yield. There are other benefits, too. It's much better for the environment generally, and it's cheaper and easier for the farmers. It boils down to an education exercise... and perhaps dealing with the interests of a fertilizer industry.

A person

0.1 ton CO_2e per year average Malawian
3.3 tons CO_2e per year average Chinese person
7 tons CO_2e per year world average
15 tons CO_2e per year average U.K. inhabitant
28 tons CO_2e per year average North American
30 tons CO_2e per year average Australian

Figure 8.1 shows two ways of looking at the emissions per person for various countries: the official "direct" footprint and my estimate for the "consumption" footprint. The direct footprints include all the greenhouse gases released inside a country's borders; the consumption footprint is adjusted to take account of imports and exports, giving a total that represents all the goods and services ultimately consumed by a typical person in each country.

For the U.K., the direct average footprint of 11 tons per head goes up to about 15 tons once you include imports and international travel and shipping. I have estimated that for

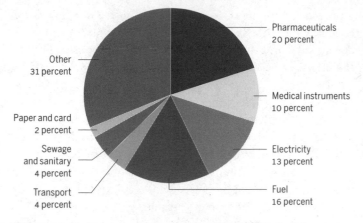

FIGURE 8.1: Emissions per person and an estimate of footprint per person.

other developed countries the same markup of about 4 tons per head seems reasonable, in which case for North America, 24 tons per head becomes 28 tons. In China the effect works in reverse. About one-third of their emissions go into exported goods, so the footprint of Chinese consumption is only about two-thirds of the emissions that physically come out of the country itself. I've estimated that a similar story applies to India but to a slightly smaller extent.

My estimates of the difference between a nation's emissions and its consumption footprint are very conservative, for one very important reason. The figures are based on the flawed assumption that the overseas production is exactly as carbon intensive as the U.K. equivalent. In other words, it assumes that if you have your washing machine, your computer and your pair of jeans made in China they will have the same embodied footprint as if they were manufactured in the U.K. We know that this is not true and there is a strong argument for using much higher figures for most imports, based on inefficient production and more polluting electricity generation in coal-dependent exporting countries such as China.

10 tons to 100 tons

A car crash

Zero g CO_2e a tiny bump that you can live with
7 tons CO_2e a write-off on an empty road
50 tons CO_2e a double write-off on a busy highway

My medium-carbon scenario involves writing off a $15,000 car on an empty road without damaging anyone or anything else. As a rule of thumb, the embedded emissions in a car are around 500 g CO_2e per $1 of value.

At the high end of my scale you write off two cars, each worth $22,500, and cause a 10-mile backup of crawling traffic across three lanes of a highway for 2 hours. If the line has frequent stops and starts, the 6,000 or so cars involved will be unable to turn their engines off and will each emit perhaps 5 kg (11 lbs.) CO_2e more from their exhaust pipes than they would have if the road had been clear. That adds up to a further 30 tons of emissions—which is more than the crash itself.

My sums have not taken account of the footprint of a whole string of other consequences of the crash: the extra burden on the emergency and health services, congestion on surrounding roads, wear and tear on cars, to name but a few. If complex surgery were involved for one or more injured drivers, that could boost the total footprint significantly (see page 129).

FIGURE 9.1: The human impact of a car crash in which two $22,500 cars are written off, a 50-year-old dies, and there is a 10-mile motorway queue lasting 2 hours. The figures exclude the indirect effects on family and friends.

It is also interesting to look at the human impact of the crash. Let's say for the sake of argument that one 50-year-old person dies but everyone else is more or less fine. We could say, a little simplistically, that the human impact has been the loss of 30 person-years of life (plus a massive impact on the lives of friends, families, and colleagues).[1] As for the broader impact, if the 6,000 cars each had a typical 1.6 occupants, about 20,000 person-hours will be spent in the living hell of the line. That's a bit more than most of us spend awake over a 3-year period, but to stick with round numbers, let's say that the human impact is 3 person-years of life lost. Finally, there are the victims of climate change. The extent to which people around the world will be affected by the release of 50 tons CO_2e is impossible to quantify, but the exercise of trying might still help us gain perspective. In an earlier chapter of this book ("A quick guide to carbon and carbon footprints") I did some rather flaky calculations to arrive at a less-than-robust figure of one life lost for every 150 tons CO_2e emitted. If this bears any resemblance to the truth, then the footprint of the crash and the line together will trigger about one-third of a life lost—say, the loss of another 25 person-years.

A new car

6 tons CO_2e Smart car, basic specification
17 tons CO_2e Ford Taurus
35 tons CO_2e Land Rover Discovery, top of the range
> So a new gas guzzler could eat up three and a half years' worth of 10-ton living before you even drive it off the lot. (It's not as much as this if you trade in your old car for resale.)

The carbon footprint of a car is immensely complex. Ores have to be dug out of the ground and the metals extracted. These have to be turned into parts. Other components have to be brought together: rubber tires, plastic dashboards, paint, and so on. All of this involves transporting things around the world. The whole lot then has to be assembled, and every stage in the process requires energy. The companies that make cars have offices and other infrastructure with their own carbon footprints, which we need to somehow allocate proportionately to the cars that are made. When you stop to think about it, the manufacture of a car causes ripples that go right throughout the economy. To give just one simple example among millions, the assembly plant uses phones, and they in turn had to be manufactured, along with the phone lines that transmit the calls. It goes on and on forever.

Attempts to capture all these stages by adding them up individually (the so-called process-based approach to carbon footprinting) are doomed from the outset to result in an underestimate, because the task is just too big. Luckily there's an alternative in the form of the input–output method (see page 191). This approach takes account of all these infinite ripples even if it does rely very heavily on the law of averages. It can give us clues as to the footprint of a car per unit of monetary value and also tell us a bit about how that footprint comes about.

Model	Price (thousands of dollars)	Embodied emissions (tons CO_2e)	Embodied emissions per mile over 100,000 miles (grams CO_2e)	Embodied emissions per mile over 200,000 miles (grams CO_2e)
Citroen C1	12–15	5–7	50–70	25–35
Prius hatchback	28–35	13–17	130–170	65–85
Ford Contour	29–43	14–20	140–200	70–100
Land Rover Discovery	49–74	23–35	230–350	135–175

TABLE 9.1: The carbon footprint of cars themselves per mile.

The input–output approach suggests that an automobile might have a footprint of 480 kg (1,050 lbs.) per $1,000 that you spend on it. Figure 9.2 shows how this breaks down. The gas and electricity used by the industry itself, including all the component manufacturers as well as the assembly plant, account for less than 12 percent of the total. The complexity of the pie illustrates just how far and wide the rest of the footprint is dispersed. There is only room to put labels on the biggest slices. This is just the beginning, though. Remember that behind each piece of pie are all the complex supply chains that lie behind that industry in turn. No wonder the process-based approach didn't stand a chance.

The upshot is that the embodied emissions of a car typically rival the exhaust pipe emissions over its entire lifetime. Per mile, the emissions from the manufacture of a top-of-the-range Land Rover Discovery that ends up being scrapped after 100,000 miles may be four times as much as comes out the exhaust pipe of my Citroen C1. I have seen plenty of analyses of the question of whether it is a lower-carbon option to keep or to scrap your old car. These almost always rely on process-based approaches and therefore underestimate the embodied energy and conclude that you should replace your car far too readily.

My family has a 12-year-old Volvo that does a disgraceful 29 miles to the gallon. If we let it go, it would be scrapped, so the embodied emissions can be considered to be written

FIGURE 9.2: The carbon footprint of a car. Gas and electricity used by vehicle manufacturers themselves account for 12 percent between them. Each of the "other" slices could be broken down again into yet another pie with a similar story. For example, the footprint of the "plastic products" slice comes partly from gas and electricity used by that industry and partly from everything else across the economy that the plastics industry spends money on. "NEC" stands for "not elsewhere classified."

off. Keeping it for the very few journeys that require a big car enables us to do almost all our driving in a little C1. So having a big banger for a second car is a lower-carbon option than having just one new large car.

Generally speaking, then, it makes sense to keep your old car for as long as it is reliable, unless you are doing high mileage or the fuel consumption is ridiculously poor. You can of course boost the life of the car by looking after it. Table 9.1 shows how much lower the total emissions per mile can be if your car lasts twice as long.

When you do eventually replace your car, do so with a light, simple, and fuel-efficient model: that way you'll be limiting both the manufacturing and the exhaust pipe emissions.

(see Driving 1 mile, page 66). But before you buy, have a look into car-share schemes: you may save lots of money as well as reduce the number of cars that need to be produced. Or if you think you need a car as a personality extension, consider getting a solar roof or a wind turbine instead (see pages 130 and 144).

A wind turbine

30 tons CO_2e a 15-kilowatt turbine, installed
500 tons CO_2e net savings over a 20-year lifetime
> WARNING: This section contains payback calculations that will interest some more than others.

Let's have a look at how a turbine might stack up in both cash and carbon terms. A 15-kilowatt turbine is at the biggest end of the micro-renewables spectrum. With a 9-meter (30-foot) diameter and a pole as high as a four-storey house, this is the most efficient form of micro wind turbine, and the sort of thing you could install only if you had plenty of space and money.

According to Chris Goodall's estimates,[2] a 15-kilowatt turbine (that's the maximum output, which is a long way above what the device typically generates), costing $61,500 to purchase and a further $14,500 to install, is capable of delivering 25,000 kilowatt-hours of electricity each year if placed on a suitably windy site.

I don't know of any credible studies of emissions of producing and installing turbines, and input–output analysis is likely to be crude even by its own standards in this rapidly changing industry. For this reason, my estimates here are going to be even more broad than usual. However, it is worth having a go. If turbine manufacture is about as carbon intensive per dollar of product as other generators and electrical motors, which seems a reasonable assumption, the carbon intensity of manufacture will be around 430 kg (950 lbs.) CO_2e per $1,000 of value. Installation is probably about as carbon intensive as typical construction, at around 250 kg (550 lbs.) per $1,000. That makes the footprint of the installed turbine 30 tons CO_2e.

The carbon savings from generation depend on the carbon intensity of the electricity that you're replacing. Let's assume that your generation replaces the coal-fueled part of the country's energy mix. In other words, let's say that rather than replacing typical grid electricity, which comes from a mix of coal, gas, oil, and renewables, the effect of your turbine is to reduce the use of coal-fired power stations. That's reasonable because coal is the least preferable source in the electricity mix (see page 56). In this case the carbon savings is roughly 1 kg (2.2 lbs.) per kilowatt-hour, so you save 25 tons per year and pay back the embodied carbon in just 14 months—a great start.

So how about financial payback? The U.K. government has recently introduced a "feed-in tariff" that pays 24p (37 cents) per unit on top of all the money owners save on their own fuel bill and from selling surplus back to the grid at approximately 5p (8 cents) per unit.[3] With all this taken into account, they would get back £7,250 ($10,900) per year on their investment. That pays back in about 6 years if you forget what else might have been done with the up-front money. However, most economists would apply a "discount rate" (see page 131) to take account of the fact that it's usually more useful to have money right now than to have to wait for it. If you apply a 10 percent discount rate (in other words you'd value a firm promise to be given $1,000 in a year's time exactly as highly as an offer to receive $900 right now), the payback period becomes 8 years, but it still makes good financial sense. And if you care about the carbon savings for their own sake, it looks like a fantastic move. The carbon investment pays back in just over a year, and every year after that is a 25-ton carbon saving. (Please don't forget that all these sums rely on your wind turbine having a favorable location.)

So, at face value, the turbine looks like a great idea environmentally and, in the U.K. at least, a fairly good long-term investment economically for the person installing it. But there is a crucial perspective missing from the analysis so far. Has the government spent its money wisely? It has invested 24p (36 cents) per unit into your turbine. That works out at a massive £250 ($375) per ton of carbon saved.

My sums tell me that had the U.K. government invested its money in large offshore wind farms, instead of subsidizing smaller turbines, they would have broken even after 8 years (without discount rate) or 15 years (with a 10 percent discount rate).[4] In other words, thanks to the feed-in tariff, the micro turbine works out as a good investment for the owner but only because the U.K. government spends, and arguably wastes, so much money subsidizing it. There are carbon savings, but they are far less than could be had if the investment were put into, for example, offshore turbines.

Suppose that the wind turbine feed-in tariff were to encourage enough micro-turbine installations to reduce direct U.K. emissions by 1 percent. That would cost £1.8 billion ($2.7 billion)—in contrast with getting the same effect from macro-renewables, which would ultimately have paid back financially. In this light, although the wind-turbine feed-in tariff doesn't look like the very best way of spending government resources on climate change mitigation, we are talking about investing about 0.075 percent per year of the nation's GDP to get a 1 percent reduction in the emissions. In other words, it could be much better but it could be worse. The investment also supports an important developing technology.

There is one extra favorable way of looking at the micro wind turbine, even if it is not the single best way of investing money in cutting carbon. Input–output modeling has told us that it is actually quite difficult to spend money without having a negative carbon impact (see Spending $1, page 60). So if the feed-in tariff encourages people to spend their money on a carbon-reducing technology such as a wind turbine, rather than on carbon-producing goods and services such as a car or a series of overseas holidays, then the reductions in emissions will be greater than my simple sums above have suggested. (See also Photovoltaic panels, page 130.)

A house

80 tons CO_2e

> That's equivalent to five brand-new family cars, eight years of 10-ton living, or 24 economy-class trips to Los Angeles to Barcelona.

This figure is for the construction of a brand-new bungalow with two bedrooms upstairs and a living room, a dining room, and a kitchen downstairs. Figure 9.3 shows the footprint of the materials. These numbers come from a study I was involved in for Historic Scotland.[5] We looked at the climate change implications of various options for a traditional bungalow in Dumfries: leave it as it is, refurbish, or knock it down and build a new one to various different building codes. We looked at the climate change impact over a 100-year period, taking into account the embodied emissions in the construction and maintenance as well as living in the building.

The worst option by far was to do nothing and leave the old house leaking energy like a sieve. Knocking down and starting again worked out at about 80 tons CO_2e whether you built to 2008 Scottish building regulations or to the much more stringent and more expensive Code for Sustainable Homes Level 5 that demanded "carbon neutrality." For all the new-build options, the up-front emissions from construction work are paid back by savings from better energy efficiency of the house in 15 to 20 years.

However, the winning option was to refurbish the old house, because the carbon investment of doing this was just 8 tons CO_2e compared with 80 tons, and even the

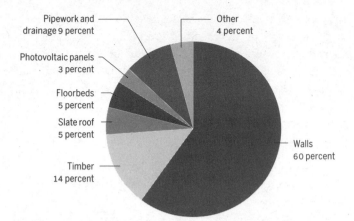

FIGURE 9.3: The embodied emissions in building a carbon-neutral two-bedroom bungalow. The photovoltaic panels struggle to justify themselves financially.

highest-specification home could not catch up this advantage over the 100-year period. Once cost was taken into account, refurbishment became dramatically the most practical and attractive option, too.

If this one study is representative, and I suspect that it is, the message for the construction industry is clear. Investment in the very highest levels of energy efficiency for new homes is, even at its best, an extremely costly way of saving carbon. Investing in improvements to existing homes is a dramatically more cost-effective approach.

100 tons to 1 million tons

Having a child

100 tons CO_2e a carbon-conscious child
373 tons CO_2e average in U.K.
688 tons CO_2e average in North America
2,000+ tons CO_2e high-impact offspring
> So if you have two typical children, that's 746 tons over
their lifetimes.

Unless you will ever contemplate lighting a forest fire (see page 161), the decision to reproduce is probably the biggest carbon choice you will ever make. The more of us there are, the greater the pressure on the world's resources.

I'm not saying you shouldn't have children. If you are someone who believes that God has told you to go and have ten of them, I am not even saying that you are wrong about that. All I'm saying is that according to my sums one of the consequences will be a few thousand tons of carbon emissions.

The average U.S. and U.K. lifetime footprints are based on your child living to the U.K. life expectancy of 79 years. I have assumed that during that time he or she will lead a nationally typical lifestyle in carbon terms and make average demands on public services. I have also assumed that throughout his or her life the average carbon footprint of a person living in the U.K. will decrease by 3.9 percent each year. This is the same annual reduction that is required if the U.K. is to meet its target of cutting emissions by 80 percent by 2050 (at which point your child will be roughly halfway through his or her life).[1]

At the high end of the scale are children who, even after you have done your best to encourage sustainability values, then go on to lead high-carbon lives. At the low end of my scale are children who grow up with carbon priorities embedded in their lifestyle and are serious about reducing emissions where they can.

All my scenarios assume that the child is living in the developed world (the numbers would be much lower in developing countries). For simplicity's sake I have not taken into account the footprint of your child's own offspring.

Deciding whether or not to have children is one thing, but the Optimum Population Trust[2] estimates that 40 percent of all pregnancies worldwide are unintended and that offering family planning in developing countries saves carbon at a rate of $6 per ton.

A swimming pool

400 tons CO_2e per year

> That's the same as 40 people living the 10-ton lifestyle or just over the expected lifetime footprint of a child born in the U.K. today.

The figures here are for a community swimming pool with a spa, costing $37 million[3] to build They are based on a real study carried out for a pool in a small town in Scotland. The study concluded that the pool in question caused a massive 17 kg (38 lbs.) CO_2e per visitor, around 30 percent of which could be prevented just through simple improvements in efficiency.

As Figure 10.1 shows, most of the pool's gas was consumed in the process of heating the water. Electricity was used mainly for pumps, air extraction, and lighting. Most visitors traveled a fair distance by car to get there, and that accounted for 20 percent of the footprint. Note that the water itself was barely significant.

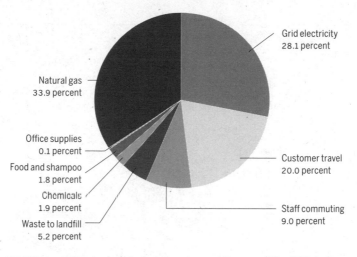

Grid electricity
28.1 percent

Natural gas
33.9 percent

Office supplies
0.1 percent

Food and shampoo
1.8 percent

Chemicals
1.9 percent

Waste to landfill
5.2 percent

Customer travel
20.0 percent

Staff commuting
9.0 percent

FIGURE 10.1: The footprint of a swimming pool in a small Scottish town.

Overall, do the high figures here mean we should all stop swimming? There are at least a couple of things to bear in mind before making that leap: the huge social benefits of swimming pools and the fact that an efficient and busy pool in a bigger town could perhaps cut the footprint per swim to one-quarter of the numbers here. Unless you actually believe your local pool should close (in which case there is an argument for avoiding going there), remember that when you swim in it, you hardly alter its footprint at all: you just put it to better use. Nonetheless, swimming remains a surprisingly high-carbon way to take exercise.

A hectare (2.5 acres) of deforestation

500 tons CO_2e

> That's equivalent to an average car driving 700,000 miles—28 times around the world.

A hectare is 100 m × 100 m (330 ft. × 330 ft.), so there are 100 of them in a square kilometer and about 260 of them in a square mile. Globally we are cutting down or burning about

	Country	Total forest area
1	Russian Federation	3,122,755 sq. mi.
2	Brazil	1,844,402 sq. mi.
3	Canada	1,197,434 sq. mi.
4	United States of America	1,170,223 sq. mi.
5	China	761,740 sq. mi.
6	Australia	631,964 sq. mi.
7	Democratic Republic of the Congo	515,871 sq. mi.
8	Indonesia	341,681 sq. mi.
9	Peru	265,414 sq. mi.
10	India	261,395 sq. mi.

TABLE 10.1: Top ten countries by total forest area.

32 million acres of rainforest per year. That's about half the land area of the U.K. The result is about 9 billion tons CO_2e or 17 percent of all man-made emissions.[4]

Most of this total (about 22 million acres or 6 billion tons) involves clearing forest to make way for livestock and other agriculture. One estimate is that 20 to 25 percent of rainforest loss is due to cattle grazing, 35 to 45 percent to small holdings, 15 to 20 percent to intensive agriculture, 10 to 15 percent to logging, and perhaps 5 percent to other causes such as urbanization, mining, roads, and other infrastructure.[5]

Anything that increases the land we need for agriculture drives deforestation. Included in this list are high-meat diets, cut flowers, and biofuel crops. In Brazil, where deforestation accounts for 70 percent of emissions, rates had been falling since 2004 until a spike in beef and soy prices brought on a further increase.

Halting deforestation is potentially one of the easiest climate change wins, if only we can find the mechanism. Brazil has pledged to cut its levels by 80 percent over the next decade. That is a big win. The Amazon Fund[6] pays farmers to hang on to their trees. It works out at just $4.5 per ton of carbon saved.[7] What a bargain! It is also fantastic for biodiversity. The Norwegian government has pledged $1 billion to support this. Why don't other governments get into this kind of thing?

The Boreal forest accounts for around a third of the world's forests, with the Canadian part alone storing an estimated 186 billion tons CO_2e; that's nearly four years' worth of humankind's greenhouse gas footprint. The Boreal's role in our climate is so significant that when it is summer in the Northern Hemisphere and the trees are growing at their fastest, global CO_2 concentrations actually come down slightly.

According to the Food and Agriculture Organization of the United Nations, Canada and the U.S. rank third and fourth in terms of total area covered.[8]

A space shuttle flight

At least 4,600 tons CO_2e

> That's the same as two bananas for everyone in Canada or half a sheet of recycled toilet paper for everyone in the whole world.

A NASA space shuttle used to burn through 106 tons of hydrogen in its external tank (the big one that would fall off after a couple of minutes and disintegrate before hitting the ground) and 527 tons of extra high-energy solid fuel in each of the two booster rockets.[9]

My carbon estimate is conservative for a variety of reasons. I have made the assumption that the process of creating the hydrogen and solid fuel using energy from fossil fuels has been 80 percent efficient. In other words I've assumed that four-fifths of the energy in the fossil fuel is transferred into the shuttle fuel. That is about as efficient as hydrogen generation ever gets, and frankly I would be surprised if energy efficiency was NASA's number one priority. I would be even more surprised if the manufacturing of the solid fuel was that efficient.

Much more significantly, it might have been reasonable to add on a large chunk of footprint of NASA itself. Richard Feynman, the Nobel Prize–winning physicist who helped to investigate the *Challenger* disaster, describes the shuttle project as NASA's somewhat unjustifiable *raison d'être* after the lunar landings.[10]

My third major omission is that I haven't included any kind of weighting factor to take account of the high altitude at which the emissions are released. When it is actually in the air, the shuttle burns a different type of fuel from a normal plane, and it releases different gases. It won't be the same story as for normal aircraft, since the gases and altitudes are both different. Somewhere along the line the water from burning hydrogen probably causes some contrails, but that's about all I can say. I would love to hear from anyone who knows more about this.

Finally, I haven't bothered to factor in the embodied energy of the shuttle itself. Since each shuttle (apart from *Challenger*, which crashed after ten trips) was reused around thirty times, I think the emissions of manufacture would turn out to be very small compared with the fuel burn.

Space tourism is not a low-carbon option.

A university

72,000 tons CO_2e per year

> So that's about 5.5 tons for each staff member and student.

The figures here are based on a study for Lancaster University. It included just about everything you can think of: gas, electricity, commuting, business travel, and everything the university buys, right down to the paper clips. It even included intangibles such as software and banking services. It didn't include everyday food because the university only caters for special events.

As Figure 10.2 shows, gas and electricity between them accounted for 45 percent of the total. Staff air travel came in

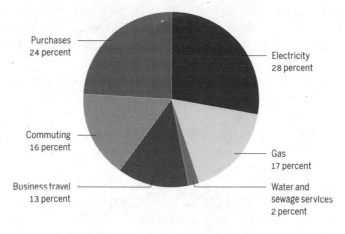

Purchases
24 percent

Electricity
28 percent

Commuting
16 percent

Gas
17 percent

Business travel
13 percent

Water and
sewage services
2 percent

FIGURE 10.2: The carbon footprint of Lancaster University.

at 10 percent, and staff and student car travel came in at about 7 percent each. Everything else that the university buys made up the remaining quarter of the footprint: IT equipment (5 percent), building maintenance (5 percent), paper based stuff (1 percent), and so on.

IT in total accounted for about 12 percent, with nearly half of that being due to the electricity consumed by computers themselves and a sixth to the power consumed by servers and other computing infrastructure, including the air conditioning to cool it all down. The remaining third was due to the embodied emissions in the equipment itself, with a little bit for services such as Internet access and software.

I'd love to be able to write about how Lancaster compares with other universities, but it wouldn't be meaningful, because no two are in the same situation. Lancaster is a fairly small campus university on a hill in the north of the U.K. with lots of buildings from the sixties and a few new ones. Oxford is farther south but can be bitterly cold in winter. It isn't on a hill but has to make the best of a load of ancient listed buildings. How could the two be compared? Also, I don't know of another university that has had quite such a comprehensive study carried out. They usually only look at electricity, gas, and travel. Lancaster has about one-third of the enrollment of McGill University in Montreal, and roughly a fifth of the enrollment of the University of Florida.

Lancaster University's footprint per head is about a third of the annual footprint of a typical British person. But don't forget that everyone at the university has plenty more in their footprint than the stuff covered by the university. The sums here don't take account of students and staff traveling to and from home, for example.[11]

How to reduce it? I won't go into too much detail here, except to say that at the time of writing, great ideas are coming from all kinds of places. Some are very technical (IT wizardry), some are mind-blowingly simple (change the light bulbs) and some are quirky ideas that bright people have dreamed up out of the blue in odd moments (like bits of hose pipe attached to air conditioning units to improve their efficiency). The biggest challenge is not having the ideas but putting them into action.

Educational establishments account for nearly 2 percent of the U.K.'s total footprint.

1 million tons and beyond

A volcano

1 million tons CO_2e Mount Etna in a quiet year
42 million tons CO_2e Mount Pinatubo, Philippines, 1991

If you have been a victim of the rumor, persistent in some circles, that volcanic emissions dwarf those of human activity, now is the time to be liberated. All the world's volcanoes together produce a total of about 300 million tons CO_2 per year.[1] This is well under 1 percent of the annual emissions from humankind's activities.

Nonetheless, as the figures above show, each active volcano does have a massive footprint, with a major eruption causing tens of millions of tons CO_2e. But these numbers are misleading because, alongside their warming effect, volcanic emissions also cause a cooling effect. The ash and sulfur dioxide that they throw up into the stratosphere reflect sunlight away from the Earth. Overall, the Mount Pinatubo eruption of 1991 is thought to have resulted in a net planetary cooling of 0.5°C the following year.

Over time the cooling effect fades faster than the greenhouse effect of the carbon, so the question of whether the warming effect or the cooling effect is greater is not clear-cut.

Iceland 2010

When Eyjafjallajökull erupted, it was estimated to have belched out 150,000 tons CO_2e per day. This climate change impact was offset by its effect on aviation. One relatively transparent back-of-the-envelope calculations concluded that the savings in aviation emissions from grounded flights was around 200,000 tons per day greater than emissions from the volcano itself.[2] I'd like to add two more factors into the equation: the first is the possibility of long-term changes in habits. A few days after initial eruption some colleagues of mine were involved in an international conference, and they had a choice between canceling altogether or holding the event virtually. They felt it was too short notice to set up the IT and so canceled. However, my hope is that others in analogous situations may have been nudged into breaking old habits and new low-carbon alternatives. The second effect is the possibility that flying through even low-level dust permanently damages jet engine efficiency. If this is the case, Eyjafjallajökull could haunt the aviation industry for many years.

Overall, was the Iceland incident good or bad for the climate? We don't know. My hunch that it did us a favor by helping society to associate flying with hassle rather than glamour.

The World Cup

2.8 million tons CO_2e the 2010 South Africa World Cup
> That's 6,000 space shuttle fights, three quiet years for Mount Etna, or three and a half cheeseburgers for every man, woman, and child in North America.

The headline footprint figure here comes from a study of the 2010 South Africa World Cup and includes players and their entourages traveling around, the construction of the sites, energy used at the stadiums, accommodation, and fans traveling (Figure 11.1).[3]

An estimated 1.2 million spectators saw matches live, so that's a massive carbon cost of 2.3 tons per viewing. Luckily

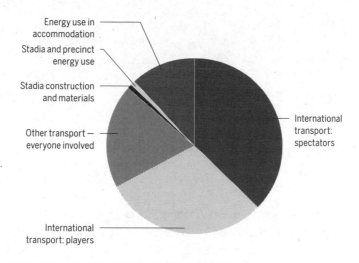

Energy use in accommodation

Stadia and precinct energy use

Stadia construction and materials

Other transport — everyone involved

International transport: spectators

International transport: players

FIGURE 11.1: A carbon footprint of the World Cup.

for the carbon credentials of the World Cup, each of the 64 matches was viewed on the television by a guesstimated 93 million people worldwide. At 2 hours per match, including intervals, extra times, penalty nail biters, and the bit where they swap shirts at the end, that adds up to a massive 12 billion fan-hours of top-quality entertainment.

If these numbers are correct, the World Cup comes in at 230 g CO_2e per fan-hour of entertainment, though of course the viewers' footprints are boosted by their own televisions. The worst scenario is that you watched alone on a 42-inch plasma screen (page 33), in which case your TV made up about half of the footprint of your viewing experience. Even if you watched 24 hours a day, 7 days a week for a whole year (that's a whole year alone without sleep in front of the plasma TV in permanent World Cup ecstasy), you'd clock up only a 4-ton footprint.[4]

By comparison, a U.K. Premiership match, at 820 tons, comes out at a tiny 45 g per viewer-hour (excluding all the televisions—see page 33), with over nine-tenths of the entertainment being exported around the world.

Even better is a kick-around in your local park or street. This is virtually emissions free, making it one of the best carbon bargains in this book.

The world's data centers

130 million tons CO_2e 2010
250 to 340 million tons CO_2e prediction for 2020[5]

> The footprint of the world's data centers is currently the same as one-seventh of the U.K.'s footprint, or a quarter of a percent of the global total.

Data centers are buildings packed top to bottom with computers. These computers store web pages, databases, applications, and downloads and generally make the Information Age possible. As you'd expect, they use lots of electricity (both for powering the machines they contain and for keeping them cool with air-conditioning), and as people consume ever more digital content, their already considerable carbon footprint is rising fast.

According to IT advisory company Gartner, the world's data centers currently account for one-quarter of the energy consumed around the world by the information and communication technology sector. That's around two-thirds as much as all the computers and monitors in the world. On current growth trends, however, the power draw of data centers is set to at least double over the coming decade. The precise growth rate will depend on efficiency improvements and changes to the amount of data being stored and processed.

If data centers alone account for 0.25 percent of the world's total footprint and that figure is set to rise to 0.5 percent and beyond, then digital data as a whole is looking set to climb to well over 1 percent of total emissions. Meanwhile 1 percent is about the proportion of the U.K.'s footprint accounted for by printing and paper-based publishing. The direct comparison is a bit more complex, but the point is that digital information may not be lower-carbon than the paper-based world of 20 years ago. Part of the problem is the so-called rebound effect—the idea that when something (in this case the storing and interrogation of data) becomes cheaper and more carbon efficient to do, we end up simply doing more of it so that there is no net reduction in cost or impact. Sometimes it is even the reverse.[6] Not only is global data growing incredibly fast, but so is our expectation that we can interrogate it at a moment's notice. Where we might have expected to line up at busy times in a bookshop to enquire about just the contents of its shelves, Amazon now has to meet our expectation, through its data center capacity, that even in peak times we can search the world's published materials in an instant.

Of course, if we go for digital information without ditching the paper, downloading stuff simply to print it out, we end up with the worst of all carbon worlds.

A forest fire

165 million tons CO_2e the Australian forest fires of 2009
231 million tones CO_2e the Californian forest fires of summer 2008
> Australian fires were equivalent to the total yearly footprint of more than 5 million Australians or 50 million Chinese people.

If you were to start one of these deliberately, that one strike of a match would make your footprint thousands of times greater than most people build up over their lifetimes.

My estimate for the Australian fires is based on 450,000 hectares (1,750 square miles) of forest containing 100 tons of carbon per hectare.[7] To put the number in perspective, the most recent estimate of Australia's annual footprint was 529 million tons CO_2e, so the fires added nearly one-third.[8]

For the Californian fires, mostly started by lightning, I've based my sums on 630,000 hectares and the same forest density.

Emissions from forest fires vary from year to year. In 1997–98 they are thought to have been around 2.1 billion tons.[9] Although some vegetation regrows, they are almost certainly a nasty example of a climate change positive feedback loop.

In theory, regrowth will absorb the CO_2 from the air in time, thus making the fire carbon neutral in the long term. However, it is looking increasingly likely that permanent changes in terrain take place following a fire. Furthermore, forest fires are also a major source of black carbon (see page 169).

A country

Around 1.5 million tons CO_2e per year Malawi
4 million tons CO_2e per year Iceland
610 million tons CO_2e per year Australia
810 million tons CO_2e per year France
862 million tons CO_2e per year the U.K.
890 million tons CO_2e per year Canada
980 million tons CO_2e per year Brazil
1,350 million tons CO_2e per year Germany
1,400 million tons CO_2e per year India
1,940 million tons CO_2e per year Russian Federation
4,300 million tons CO_2e per year China
8,250 million tons CO_2e per year the U.S.

The estimates here are of the footprints of national consumption in 2005. They include the footprint of goods imported

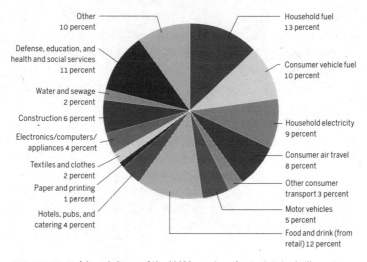

Other
10 percent

Defense, education, and
health and social services
11 percent

Water and sewage
2 percent

Construction 6 percent

Electronics/computers/
appliances 4 percent

Textiles and clothes
2 percent

Paper and printing
1 percent

Hotels, pubs, and
catering 4 percent

Household fuel
13 percent

Consumer vehicle fuel
10 percent

Household electricity
9 percent

Consumer air travel
8 percent

Other consumer
transport 3 percent

Motor vehicles
5 percent

Food and drink (from
retail) 12 percent

FIGURE 11.2: A breakdown of the U.K.'s carbon footprint, including imports but excluding exports.

from overseas but exclude goods produced in each country for export.

Let's start off by looking at the footprint of the U.K. As we saw earlier (see page 137), each person in the country is responsible for around 15 tons CO_2 per year. Because the U.K. imports more goods than it exports, the total figure for the country—862 million tons—is higher than the 700 million or so tons that usually gets reported for U.K. emissions.[10]

Taking the U.K. as an example of a fairly typical European country, let's see how all those emissions break down (Figure 11.2).

Domestic energy, which often dominates the media coverage of carbon footprints, makes up 22 percent of the total, consisting of household fuel at 13 percent and electricity at 9 percent. For most people the fuel is gas, for which 84 percent of the emissions happen in the home itself and the rest are caused during gas extraction and distribution.

Cars come in at 15 percent of the total when you add together the 10 percent caused by their fuel (extraction, distribution, and use) and the 5 percent caused by their manufacture and maintenance. As a rule of thumb, the exhaust-pipe emissions are about half the total footprint of driving (see page 67). Note that this slice of the pie doesn't include commercial vehicles, so the whole of road transport is a good deal more.

Food and drink, often underestimated, come in at 12 percent just for those groceries bought at stores. If we include all the food and drink served by hotels, pubs, cafés, schools, hospitals, and so on, we'd get to about 17 percent. If we also added in the emissions from cooking at home, traveling to the stores, and the emissions from food waste sent to landfill, the total footprint of the stuff that goes into people's mouths comes to about 20 percent of the U.K.'s footprint. It's roughly the same percentage for the world as a whole. All these numbers are without considering the impact of food demand on deforestation, which would take the U.K. total to around 30 percent.[11]

Air travel for private purposes is a staggering 8 percent[12] of the total. If you include business travel and air freight as well, flying comes in at around 12 percent of the U.K.'s footprint—much higher than the figure usually quoted. British people probably fly a bit more than other Europeans because they live on an island and because there is a lot of sea to the west, but this is still a remarkable statistic—especially when you consider that air travel is the fastest-growing major emissions source in the country.

In the pie chart, construction, at 6 percent of the total, includes domestic repairs, new houses, and all new commercial construction work. And the production of electrical goods—that is, household computers and appliances—comes in at 4 percent, almost half as much as the electricity they consume in use.

Public administration, defense, education, health care, and social services cause a significant 11 percent of emissions. A common misconception is that there is nothing we can do about this as individuals. To cut your share of these emissions, how about preventing crime; encouraging schools, universities, and businesses to manage their carbon; staying as healthy as possible; and voting with climate change in mind?

Other factors. This is my catch-all category. It contains a jumble of things, including some that might never occur to you as incurring a footprint at all. In here are bikes, brooms, Lego bricks, lipstick, legal fees, phone calls, footballs, tables, toiletries, travel insurance, jewelry, and too many other things to list individually.

So that's the U.K. How about the rest of the world? Figure 11.3 picks out a few key countries and gives their official carbon emissions (that is, the quantity of greenhouse gases released within their borders) and my estimate of their true consumption footprints (that is, with exports subtracted,

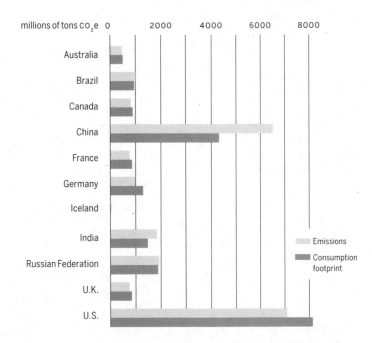

FIGURE 11.3: National emissions and consumption footprints in 2005.

emissions divided by GDP (kg CO$_2$e/$)

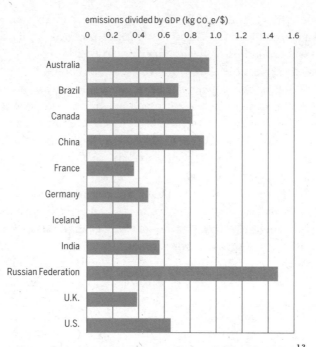

FIGURE 11.4: Emissions per dollar of GDP for a selection of nations.[13]

and imports and international travel and transport added in).
Note that the numbers shown are from the year 2005. Things
change fast, and China is now widely thought to lead the U.S.
in terms of emissions, probably having drawn level in about
2006, even though the footprint of Chinese consumption is
still just over half that of the U.S.

Even when you factor in imports and exports, however, it
isn't necessarily hugely meaningful to think of the total emissions of a country. China may have overtaken the U.S. as the
biggest emitter, for example, but it has *far* more people. Hence
emissions per person is usually a more meaningful way to
compare one nation with another—as discussed on page 193.

Yet another way of looking at a country's footprint is in
terms of emissions per unit of GDP (gross domestic product);

see Figure 11.4. This is a measure of "carbon efficiency" or "carbon intensity"—a nation's footprint relative to its economic activity. Countries with inefficient factories, and which get their electricity from dirty coal-fired power stations, rate worse on this scale. Hot countries can sometimes achieve a better rating because they don't have to spend so much on keeping warm (provided that people aren't rich enough to afford air-conditioning).

Through this GDP lens, Russia comes out worst, because of its coal-fired power stations, inefficient factories, and cold climate, with coal-dependent China and Australia following behind. Western Europe has relatively efficient factories and cleaner electricity, so countries in this region come out well—especially nuclear-powered France and renewable-powered Iceland. The U.S. comes in somewhere in the middle of the carbon-efficiency stakes.

Although we may talk about becoming more carbon efficient, it is clear that we are a very long way from being able to grow our economy without increasing our consumption footprint. Tim Jackson's recently published *Prosperity without Growth*[14] is both the most rigorous and the most accessible articulation of this uncomfortable reality that I have seen.

Ultimately, there's no avoiding the fact that a country's emissions are strongly linked to its wealth. It's hard to be rich and have a low carbon footprint (see Spending $1, page 60). Malawi is just one example of a country whose poverty ensures a low footprint. Its 14 million people have a footprint of around 100 kg each per year.

I've looked at typical footprints by country, but this doesn't always give the full picture. Sometimes the most significant differences occur *within* countries. In China, for example, hundreds of millions of people live very low-carbon lives, whereas the emerging middle class, with Western lifestyles in a less energy-efficient economy, probably have carbon footprints to dwarf those of the Australians.

A war

690 million tons CO_2e a "limited" nuclear exchange of fifty 15-kiloton[15] warheads

250 to 600 million tons CO_2e Iraq, 2003–09

> The Iraq war up until the start of 2010 probably racked up a carbon footprint roughly equivalent to the whole of the U.K. economy for between 3 and 8 months and rising.

The direct human costs of wars are so great that it might seem flippant to think about their climate change costs. But war unfortunately plays a big role in global society, so this book wouldn't be what it says on the cover without giving it a mention. Moreover, it's worth bearing in mind that even just the emissions of a war could ultimately have serious human impacts somewhere in the world.

In what was perhaps the only academic estimate of the carbon footprint of an atomic war, it was concluded that even a "small nuclear exchange" of just fifty 15-kiloton warheads would cause 690 million tons of CO_2 emissions through the burning of cities.[16] The same report also estimated that the exchange would also release 313 million tons of soot into the atmosphere, which would have a cooling effect and would therefore counter the warming for the first few years after the explosions.

But a war doesn't need to be nuclear to have a huge carbon footprint. The financial cost of the U.S. military operation in Iraq for 2003–09 has been estimated at $1.3 trillion, with a further $600 billion anticipated for the lifetime health care costs of injured troops.[17] We can use the input–output model to give a very crude estimate of the footprint of the U.S. operation, of 160 to 500 million tons CO_2e for the military activities and perhaps a further 80 million tons for the health care of troops.[18] This excludes the actual emissions from combat itself. Add on a few percent to both numbers to include the coalition forces. Also add perhaps another 1 percent for the footprint of the much more poorly resourced insurgency. Overall we might be looking at 250 to 600 million tons—roughly equivalent to between 20 percent and 60 percent of all U.S. citizens flying from L.A. to Barcelona and back. The war-and-carbon discussion starts to get distinctly uncomfortable (and methodologically just about impossible) at the point where we start factoring in the indirect emissions impact caused by the death toll and indeed the broader economic impacts of the war. In the nuclear example, the report in question estimates 17 million deaths—equivalent to around one-quarter of the U.K. population. Looked at in the starkest and simplest possible terms, if each of these people had a typical U.K. footprint, then the carbon savings of their ceasing to exist might make up for the direct emissions from the war in just a few years. In other words, mass annihilation turns out to be an effective way of curbing emissions—though of course it also defeats the object.

Black carbon

7 to 15 billion tons CO_2e per year

> That is 15 to 30 percent on top of the figure I normally quote
> for global human-made emissions.

How can this have slipped off the radar for so long?* As Dennis Clare of the World Watch Institute put it, "Black carbon, a component of soot, is a potent climate-forcing aerosol and may be the world's second-leading cause of global warming after CO_2."[19]

My low figure for black carbon's global warming impact—7 billion tons of carbon equivalent—came from the Intergovernmental Panel on Climate Change 2007 report. The higher figure of 15 billion tons came from more recent studies.[20]

Black carbon warms the world in two ways. In the atmosphere it contributes to the greenhouse effect. Down on the

---

* I inserted this entry late in the day. The implications of black carbon are not systematically integrated throughout the book. I could apologize, but then it's just an example of how we are all having to learn on the go.

ground, it turns snow and ice murky and in so doing makes it absorb more of the Sun's heat. It is thought to be a major contributor to the global reduction in ice cover, especially in the Northern Hemisphere.

Black carbon is caused by incomplete combustion; 42 percent comes from outdoor fires of one kind or another, and one-quarter comes from the burning of wood, coal, dung, peat, and any other organic stuff in homes. A further quarter comes from transport (mainly diesel), and about 10 percent comes from coal-fired power stations.

The good news about black carbon is that it lasts only a few days in the atmosphere. In other words, if we can reduce the amount we create, the benefit will be instant. Hence some experts think that reducing black carbon pollution should be a number one priority in tackling global warming. Easy wins can be made by using particulate filters on diesel engines and swapping inefficient open fires for super-efficient stoves.

The world

50 billion tons CO_2e per year

In 2007, the IPCC estimated global greenhouse gas emissions at 49 gigatons (that is, 49 billion tons) CO_2e a year and rising. That doesn't include a multiplier to take account of the extra warming effect of emissions from planes, which takes the total to 50 gigatons CO_2e. Figure 11.5 shows the emissions broken down to constituent greenhouse gases, looked at in terms of the impact over a 100-year period.

Around half the methane comes from agriculture (especially livestock at 5 percent of global emissions, but also rice cultivation at 1.5 percent and other farming). The rest of the methane comes mainly from the extraction and processing of coal, gas, and oil; from landfill (2 percent of the global total); from the treatment of used water; and from other wastes.

Nitrous oxide results mainly from the spreading of nitrogen fertilizer and manure, although there are also contributions

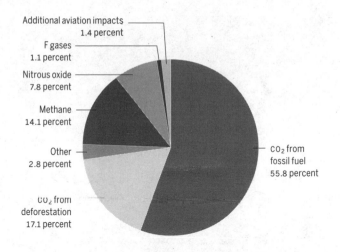

Additional aviation impacts
1.4 percent

F gases
1.1 percent

Nitrous oxide
7.8 percent

Methane
14.1 percent

Other
2.8 percent

CO_2 from
deforestation
17.1 percent

CO_2 from
fossil fuel
55.8 percent

FIGURE 11.5: The breakdown of our 50 million tons CO_2e greenhouse gas.[21]

from fuel combustion, industrial processes, and waste treatment. The F gases, at 1.1 percent, result mainly from refrigeration and air-conditioning (there are some good technologies
coming through to deal with this: see Refrigeration, page 179).
Aviation comes to 3 percent of the total once you factor in the
effect of altitude (in the pie chart the 1.4 percent sliver is just
the additional bit to take account of high-altitude effects).

Although there is a convention to look at the impact of different gases over a 100-year time span (that is, up until 2110),
this is in fact somewhat arbitrary. Many people believe that
climate change will bite much sooner than that, so there may
be a case for considering the impact over shorter timescales
as well. This changes the relative impact of the different gases:
those that are powerful but short-lived become more important relative to the weak but long-lasting CO_2. If you were to
look at a 50-year timescale, for example, the non-CO_2 emissions caused by agriculture, refrigeration, and air-conditioning would immediately become roughly twice as serious.

To understand *where* in the world all these emissions come
from, it's time for some squidgy maps. The first, Figure 11.6,

FIGURE 11.6: The world according to greenhouse gas emissions (as of 2000).[22]

FIGURE 11.7: The increase in global greenhouse gas emissions (1980–2000).[23]

FIGURE 11.8: Global emissions of methane and nitrous oxide in 2000, in CO_2e.[24]

shows the size of countries in proportion to their emissions. Note that these are just the emissions that physically rise out of each country rather than the consumption footprint, so the emissions from a factory in China that makes washing machines for people elsewhere in the world is shown as belonging to China, rather than to the people who buy them. The U.S., China and Europe dominate the map. More recently, China has overtaken the U.S. as the biggest national emitter.

Figure 11.7, which focuses on emissions growth, shows how things have been changing. In this view, China, India, and the U.S. dominate, while most of Africa has been strangled to nothing. A few European countries have vanished, too, showing that their emissions were static or declining.

Figure 11.8 shows just methane and nitrous oxide, and in doing so it emphasizes agricultural emissions. In this view China, India, and South America dominate. North America, although hardly looking lean, is at its smallest. This is by far the most imposing view of Africa out of the three, although it still looks emaciated. A bit more fertilizer application in Africa could actually be helpful.

Burning the world's fossil fuel reserves

2.5 trillion tons CO_2e

> That's 50 years of current global emissions.

The exact figure depends on just how big you think our reserves are. The numbers in Table 11.1 are based on "proven" reserves, as of 2006, but nobody really knows for sure how much is down there. Because fossil fuels account for only a little over half of total global emissions, even this conservative estimate of our reserves means that we have enough fuel left to allow us to keep on belching out carbon at our current rate until roughly the end of the century.

In other words, regardless of the precise amount of fossil fuel left in the ground, it's clear that there's more than enough to push us into climate meltdown if we were to burn it all—and judging by our current mindset it does look as though we are on course to do just that. For this *not* to happen, we will need to achieve a situation in which although there is fuel in the ground waiting to be extracted and burned, the countries and businesses that own the rights to it are simply content to leave it there for all time.

For me this is an uncomfortable perspective because it is hard to imagine Russia, China, the U.S., Saudi Arabia, Exxon, Shell, BP or anyone else simply being happy to leave their valuable assets down there in the ground. One solution, albeit a long way off, might be to devalue those assets by making renewable energy even more abundant.

Oil spills

At the time of writing, the BP Horizon Deepwater oil spill is estimated to have topped out at 300,000 to 600,000 tons of oil, assuming BP's claims to have stopped the flow prove to be well founded.[25] If (and it's a big if) the eventual fate of all the oil is to degrade into carbon dioxide, the result would be roughly one to two million tons CO_2e. That is a lot, but in this case, the climate change implications look small compared with the wider environmental impact of the disaster.

For comparison with other big spills in the history of oil, Saddam Hussein's deliberate 1991 gulf war spill is estimated at 270,000 to 820,000 tons, and the 1910–11 Lakeview Gusher in California is estimated at 1,230,000 tons.

Fuel	Billions of tons oil equivalent	Billions of tons CO_2e from burning
Coal	463	1,470
Oil	165	530
Gas	163	520
TOTAL	791	2,520

TABLE 11.1: Proven fossil fuel reserves.[26]

More about food

FOOD DESERVES a chapter of its own because in terms of carbon footprints it is such an important but poorly understood area. The various food entries early in this book have covered many of the key points, but this chapter briefly pulls together all the main issues to give a sense of the food sector overall. Most of the estimates in this section have come from my work over several years for Booths, a U.K. supermarket chain. Some of the specifics of the food you buy will depend on where you live in the world, but the principles are largely transferable.

As we saw earlier, in the developed world, the food we buy adds up to around 3 tons CO_2e per year.[1] In the U.K. that's 170 million tons CO_2e per year or 20 percent of the average annual total footprint and nearly as much as household fuel and electricity put together. In North America, it is likely to be a lower proportion, since the average per capita is nearly double the U.K. figure. If you factor the effect of deforestation, the footprint of our food goes up again to a staggering 30 percent of the U.K. total. Interestingly, food is also a very expensive part of our footprint. If you want to trash the planet, buying the wrong food or wasting what you buy is a far more expensive way of going about it than leaving the lights on or turning up the thermostat.

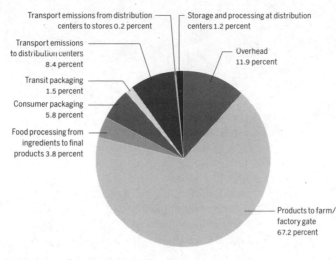

FIGURE 12.1: Total footprint of Booths products and supply chains: 226,000 tons CO_2e.

How the footprint of food breaks down

Figure 12.1 shows an estimated breakdown of the footprint of food at the point when it leaves a Booths. This is just a best estimate. As the chart shows, two-thirds of the impact is on the farm. Transport is a big deal for some products but not most. The supermarket's own operations make up about one-ninth of the total picture.

Farms

Whereas CO_2 is the dominant greenhouse gas overall, it accounts for only 11 percent of agricultural emissions.[2] The rest is nitrous oxide (53 percent) and methane (36 percent). Nitrous oxide is 296 times more potent per pound than CO_2 as a climate-change gas, and on farms it results mainly from the use of fertilizer but also from cattle pee, especially if there is excessive protein in their diet, and from the burning of biomass and fuel.[3] Methane, which is 25 times more potent than CO_2, is mainly emitted by cows and sheep when they belch. Some is also emitted from silage. The CO_2 comes from machinery but also from the heating of greenhouses to grow crops out of season or in countries that just don't have the right climate.

Transport

The first thing to say about transport emissions is that, for all the talk that we hear about food miles, they are not the most pivotal thing to think about. At Booths, over one-quarter of the transport footprint comes from the very small amount of air freight in their supply chains—typically used for expensive items that perish quickly. Conversely, most of their food miles are by ship (partly because the U.K. is an island), but because ships can carry food around the world around 100 times more efficiently than planes, they account for less than 1 percent of Booths' total footprint. The message here is that it is OK to eat apples, oranges, bananas, or whatever you like from anywhere in the world, as long as it has not been on a plane or thousands of miles by road. Road miles are roughly as carbon intensive as air miles, but in the U.K. the distances involved tend not to be too bad, whereas in North America they can be thousands of miles. Booths is a regional supermarket with just one warehouse, so their own distribution is not a big carbon deal, and they have been working hard on further improvements.

Meat and dairy

Food from animals turns out to be more carbon intensive (remember, this is my shorthand for *greenhouse gas intensive*) than food from plants, simply because animals are inefficient devices for producing food. They eat plants and then spend their lives wasting most of the energy from them on things such as walking around and keeping warm. It is a far more efficient process for humans to eat plants directly, so that all the plant energy can go directly to us. Beef and lamb are doubly high in carbon because they are belching ruminants. Chicken is a bit better because, to put it bluntly, they don't live as long, so they don't get so much opportunity to waste the energy in their feed.

Dairy has all the same problems of ruminant meat production, so there is little point in switching from beef to cheese. A kilo (2.2 pounds) of cheese comes in at around 13 kg CO_2e, compared with around 17 kg for beef. Milk comes in at around 1.3 kg per liter or quart.

Hothouses
Protected crops can have just as high an impact as air-freighted foods. It takes a huge amount of energy to keep a greenhouse warm enough to grow tomatoes during the winter (see page 99).

Packaging
This topic needs to be kept in perspective. It's only about 6 percent of what you should be considering as you shop. And at its best it serves a purpose, helping food to stay fresh and letting you know what you are buying. Indeed, a simple bag can dramatically improve the shelf life of some fresh foods.

At Booths we found that no single material dominated the packaging footprint, and there were some surprises.

> Paper and cardboard are often more carbon intensive than plastic packaging, mainly because making paper is so energy intensive but also because it emits methane if it ends up in landfill.

> Plastic is environmentally nasty as either landfill or litter because it hangs around for so long. However, it is typically not quite as energy intensive to produce as card packaging and has the advantage, from a purely carbon perspective, that when you put it in landfill, you are just sending those hydrocarbons back into the ground where they came from for long-term storage. In the days when supermarkets routinely

gave out disposable plastic bags, they accounted for around one-thousandth of the footprint of a typical shopping trip. Biodegradable plastic can be a well-intentioned nightmare, clogging up recycling processes, with the potential to ruin a whole batch. In landfill it rots, emitting methane.[4]

> Glass is energy intensive to make (or recycle), and its weight adds to the transport footprint. Cans of beer are better than bottles, as are cartons or boxes of wine. Incidentally, bottles are absolutely no better for storing wine than the more climate-friendly alternatives.

> Steel and aluminum are carbon-intensive stuff, but you don't need a great weight of them, and they're easy to recycle. It takes only about one-tenth of the energy to recycle aluminum compared with extracting it from ore in the ground.

Food waste

In the developed world we are thought to waste about one-quarter of the edible food we buy.[5] This figure depends partly on your definition of what was edible in the first place. Do you think of the potato skin as just packaging, or do you think of it as the tastiest and most nutritious bit? Whatever your definition, a huge and expensive proportion of our food gets left on plates, is allowed to go off in the fridge, isn't scraped out of the pan properly or isn't picked off the carcass. It is slightly better to compost waste food than to throw it into landfill, but it doesn't get you away from the main issue that the carbon footprint of that food has been needlessly incurred.

Refrigeration

Fridges use electricity, and it takes energy to make them in the first place. On top of that is the problem that traditionally they have relied on the use of refrigerant gases that have a global warming potential several thousand times that of CO_2. This stuff tends to leak out of large commercial fridges, which need topping up regularly. At Booths, this leakage from within the stores and warehouse accounted for around 3 percent of the total footprint. And refrigeration accounted for about half of all electricity usage in stores. When all considerations are taken into account, refrigeration probably accounts for around 6 percent of the footprint of supermarket food.

There are huge strides being made in cooling technologies. These include the use of other gases with dramatically lower global warming potential,[6] the reuse of spare heat to warm the stores, and the use of underground cooling pipes. Booths is starting to employ CO_2-based refrigeration systems (thereby almost eliminating the climate change impact of gas leaks) and expects to have replaced almost all its fridges with these within a decade. The company is also reusing the heat in their newest stores. Thanks to these kinds of approaches, we can expect the footprint of commercial refrigeration to fall dramatically. In the meantime, do not let it put you off your fresh, chilled produce.

Low-carbon food tips

The following is a quick summary of the various steps you can take to reduce the carbon footprint of your diet—and the type of savings you can expect.

Eat what you buy. Ask people how much they would like before you serve them. Eat the skins. Clean the plates, pick the carcass. Save the leftovers. Check what needs to be eaten when you plan your menus. Keep vegetables in the fridge if you can. Rotate the contents of your cupboards so that old stuff is at the front. Eradicating waste is worth a 25 percent savings for the average shopper.[7]

Reduce meat and dairy. I'm not saying go vegan any more than I'd say never drive. But there is no dodging the fact that meat and dairy are key areas. By reducing our consumption of these food types, many of us will live a bit longer and save money as well as reducing our emissions. The vegetarians and vegans I know don't consider it a hardship. Sensible reductions in meat and dairy without needing to go vegetarian are probably worth another 25 percent savings on a typical diet.

Go seasonal, avoiding hothouses and air freight. Local, seasonal produce is best of all, but shipping is fine. As a guide, if something has a short shelf life and isn't in season where you live, it will probably have had to go in a hothouse or on a plane. In the U.K., Canada, and more northern parts of the U.S., in January, examples are lettuce, asparagus, tomatoes, strawberries, and most cut flowers. Apples, oranges, and bananas, by contrast, almost always go on boats. Adopting this tip religiously can probably deliver a 10 percent savings on a typical diet.

For more specific information, try the following:

> The Eat Well Guide to seasonal food in different U.S. states and Canadian provinces: www.eatwellguide.org/i. php?id=Seasonalfoodguides

> Epicurious's interactive seasonal recipe map of the U.S.: www. epicurious.com/articlesguides/seasonalcooking/farmtotable/ seasonalingredientmap

> Food Down the Road's simple chart showing Canada's seasonal foods by month: www.fooddowntheroad.ca/online/ seasonalfoodchart.php

Avoid low-yield varieties. Cherry tomatoes and baby corn are classic examples. Estimated savings: 3 percent.

Avoid excessive packaging. Some packaging serves a valid purpose in keeping food fresh. But a metal dish inside plastic trays inside a plastic bag within a cardboard box is probably excessive. Worth around 3 to 5 percent.

Recycle your packaging. Worth 2 to 3 percent.

Help the store reduce waste. Always take from the front of the shelf so that the stock can be rotated. Handle food with care. Buy the reduced-price items when you can, but don't hang around waiting for them to be reduced. Worth perhaps a 1 percent savings.

Buy misshapen fruit and vegetables. Stimulate demand for the huge quantities of produce that get thrown away just because of their shape. The savings are hard to quantify, but perhaps 1 percent.

Lower-carbon cooking. Use a pan lid whenever you can. Remember that water boils at the same temperature however much heat you apply, so for cooking food, a gentle boil is just as fast as a furious one. Use a microwave when appropriate. Perhaps a 5 percent savings.

Incredible! The savings here add up to about 75 percent. Sadly the math doesn't work out quite like that because some of these points overlap. If you do them all, they work out to more like a 60 percent savings—still a remarkable amount.

Further information

Assumptions revisited

I started out with three assumptions:

> Climate change is a big deal.
> It is caused by humans.
> We can do something about it.

This book isn't really about those assumptions, but this section is for anyone who is still unsure. The human capacity for collective denial is an amazing phenomenon to watch. If that is where you are right now, I'm not too hopeful that I can shift you.

Is climate change a big deal and caused by humans?

At the end of the day we all have to make up our own minds. I can't go over the scientific arguments in detail here, and even if I did I'd just be one more voice for you to sift through. But I will briefly go through how I came to make up my mind.

None of us really knows for sure what climate change is going to mean for us in the coming decades. The science is hideously complex and uncertain. The media still report a full spectrum of arguments. It's a confusing picture for the lay-person. What basis can we have for knowing whether a news article, a TV program, or a book is credible?

A key question in this context is *how can we work out whom to trust?* I meet plenty of people who have understandably given up trusting anyone over climate change. But it is possible to do a lot better than that. This is how I make up my own mind about a report or a piece of research:

1. I look at the argument itself and see if the logic makes sense at face value.

2. I look at the competence of the source.

3. I look at the resources and information that it had at its disposal.

4. Critically, I try to understand the motivations—political, financial, and psychological. How strong was the dedication to truth? Who funded the research, and what did those funders want? Who wanted what from their careers, and what influence might this have had? What was the psychological readiness of the source to accept and report on different findings that might emerge?

These are the questions I have been asking about skeptics' arguments. They can sometimes pass the first test, but every single one of them fails at least one of the final three.

A few years back, just before I reoriented my working life toward addressing climate change, I thought I'd better double-check that the whole thing wasn't a storm in a teacup. I didn't want to go to a whole lot of trouble for nothing. I knew my family was going to have to put up with my hardly earning anything for a year or two while I learned a new trade.

A good friend of mine had raved about Bjørn Lomborg's book *The Skeptical Environmentalist.* "Mike," he said, "I've read this book and it's rearranged my thinking." It's a thick and persuasively written tome with some 2,000 academic references. It makes the claim that we can all afford to chill out about climate change and we would do better to invest the money elsewhere. Lomborg further asserts that the climate change worriers are psychologically wedded to a doom-and-gloom position on life. To me, that last point hit a nerve.

It was an important challenge to address. I thought, "Perhaps he's right! Maybe I should ask myself if this applies to me?" I didn't want the experience of realizing in years to come that the only reason I've done all this stuff about climate change is because of some unhealthy personal hang-up. At the very least I felt that the mainstream scientific community should have a blisteringly clear response to Lomborg, and it was disquieting that I couldn't readily find one.

I sat down to spend about a week with Lomborg's work. I picked into some of his arguments in detail and before long found that even from my distant position I could see several clear misrepresentations of science. Then I found that his book had never been appropriately peer-reviewed. Then I started uncovering websites that detailed his errors literally in their hundreds, along with roasting dismissals of his arguments from scientists, statisticians, and economists alike. After that I started to read about Lomborg's close shaves with the Danish Commission for Scientific Dishonesty. In the end it was abundantly clear to me that the whole thing was a sham.[1] I came to a clear view, but it took detailed consideration of his work—far more than can be expected of the average person on the street.

Lomborg passed the first and third of my tests but failed the second and fourth. To this day Lomborg carries on and has a following. It is incredibly unhelpful for the world. I don't know any scientists who have any time for his position at all, although some commentators treat his work with unwarranted respect in the misguided name of "balance" or perhaps just to be polite.

In the name of open-mindedness I've looked in detail at several other "skeptics" and had a similar experience.[2]

So much for the skeptics. Let's look at the mainstream scientific community. The UN's Intergovernmental Panel on Climate Change consists of around 2,500 scientists. The skeptics point out that there may be potential for group-think and mass hysteria. These are warnings that should be taken seriously. Furthermore, there have been occasional errors in the IPCC's work, and even the hint of the odd deliberate misrepresentation. However, the standard of integrity that is demanded of the climate change believers is on a different plane altogether from that demanded of the skeptics. Some scientists at the University of East Anglia have been in world in headline-hitting trouble for allegedly "sexing up" their work in a way that the some of the skeptics would consider quite normal. The resulting scandal, which turned out to be about not all that much, has been hugely damaging to popular understanding of climate science.

It's worth bearing in mind that it would also be possible to criticize the IPCC for its caution. Does it offer a sufficient platform for the airing of discomfort about poorly understood scientific risks? Does the level of deliberation and the need for consensus among such a wide community, some members of which have clearly been under political pressure to play things down, result in an undercooked estimation of the risks? We can't know for sure. We do know that the extent of scientific consensus is almost unanimous in affirming the first two of my assumptions.

Finally I want to note a trend that I have also picked up on among the people I know. The more scientifically minded they are and the more they have thought about the issues, the more worried they tend to be that even though we *might* almost all be fine, it is also just as likely that we'll end up frying in our billions. I talk to a lot of academics, mainly physical scientists and social scientists. In the last few weeks I've started conducting my own informal opinion poll by asking any senior academic that I meet to estimate the percentage of people in their department who think that "climate change is a big deal and is caused by humans." So far I have yet to have anyone give me a figure under 99 percent. It is an amazing phenomenon that people within the academic community, those with the most realistic and mature under-standing of how the academic process works and of how sci-entific knowledge evolves, are so clear about my first two assumptions while the wider public remains so obstinately doubtful.

Can we do something about it?

People ask me sometimes why they should bother when, even if everyone in their country cut the carbon, it would make such a small impact on world emissions. Sometimes I hear businesspeople trying out the argument that their hands are tied until governments act or until their end consumers care more. Governments say they can't move ahead of popu-lar opinion. I hear Chinese people saying that the developed world started it and is more carbon hungry, so they should start the cuts, whereas in the U.K. I hear people saying we're just a pinprick in comparison with the U.S. or the emerging Chinese middle classes.

The UN climate negotiations in Copenhagen and elsewhere have surely taught us that it isn't enough to hope that world leaders will sort things out on their own. So the question is, Where does leadership come from? My answer is that it can come from anywhere, and we need it to come from every-where at once. If the Chinese middle class wants a Western lifestyle, then Western lifestyles had better become lower carbon. Who can start that off? Anyone can. Anyone who finds a way of enjoying life more for less carbon is setting a standard for others. Anyone who chooses a lower-carbon food is helping the supermarkets to emphasize that prod-uct. Any supermarket that improves and promotes its lower-carbon range is helping its customers to enjoy low-carbon food. All of this helps the political parties to move into a low-carbon position.

If you can find a way of being happier but with a smaller footprint, you are a leader.

The cost efficiency of selected carbon-saving options

The list I give below isn't complete, but I have included it to illustrate that it is essential to pick our battles. Taking the U.K. as an example, some of the least cost-effective options on this list are receiving major government funding, while some of the best-looking options haven't yet had serious attention. There could be other well-founded reasons for this, but they aren't yet obvious to me.

It can be frustrating to see public money wasted on red herrings, apparently because the analysis simply hasn't been done. Quantified carbon and cost analysis may not be the whole story, but it is an essential part of it.

All the figures below are net costs or profits over the lifetime of the measure. They are based on a financial discount rate of 10 percent (see Photovoltaic panels, page 130). In other words, if you are promised a savings of $1,000 but have to wait a year for it, I've only called it $900. If you have to wait 2 years, I've called it $810, and so on.

> **Putting 270 mm (10-inch) attic insulation in homes that haven't got any**
> $105 net profit per ton saved. $2.80 for every $1 invested.

> **Investing in offshore and onshore wind farms**
> Just above zero. Payback in 15 years (would be 8 years if we ignore discount rates). Lifetimes of the farms vary.

> **Slowing down from 70 miles per hour to 60 miles per hour on the highway**
> Variable, but typically cost neutral even when the value of the driver's time is included. No investment costs (see page 66).

> **Pay farmers to keep their forests via the Amazon Fund or similar**
> $4.5 per ton, plus biodiversity benefits (see Deforestation, page 151).

> **Funding family planning in the developing world**
> $6 per ton according to the Optimum Population Trust (see Having a child, page 149).

> **Upgrading attic insulation to 270 mm (10 inches)**
> **where 50 mm (2 inches) currently exists**
> $7.5 per ton. This figure is the total cost, which is shared
> between government and homeowner.
> **U.K. government investing 24p (36 cents) per unit**
> **to a feed-in tariff for micro wind turbines**
> £250 ($375) per ton saved, assuming that this replaces elec-
> tricity from coal, and ignoring the embodied energy in the
> panels themselves (see Wind turbine, page 144).
> **U.K. government investing 36.5p (55 cents) per unit**
> **to a feed-in tariff for micro-photovoltaic panels**
> £360 ($540) per ton saved, assuming that this replaces elec-
> tricity from coal, and ignoring the embodied energy in the
> panels themselves (see Solar panel, page 130).
> **Building to U.K. code for sustainable homes level 6 (carbon**
> **neutral) instead of to current U.K. building regulations**
> Almost certainly very expensive (see A house, page 147).

Where the numbers come from

I hope I have already made the point clearly enough that
carbon footprinting is a long way from being an exact process, whatever anyone ever tells you or whatever numbers
you might see written on the side of products in some stores.
All my numbers are best estimates and nothing more, even
though I have reached them as carefully as I can.

I have tried to be as transparent as I can within the practical constraints of the book and my resources. Occasionally
the sources are confidential to clients of mine, but more often
it is simply too laborious to document every last detail. Nevertheless, there is a reasonable degree of transparency most of
the time, and here is a summary of my approach.

I have used a variety of different methods and sources. I
have drawn on a range of publicly available data sets and
models, from life-cycle studies and reports, and from studies I have carried out myself for businesses across different

industries. I have used models that we are developing all the time in my company, Small World Consulting.

Often I've arrived at numbers from a couple of different routes to check that the results agree with each other. I've tried to put notes and references in the text wherever possible. Occasionally, frankly, it has been more a case of putting my finger in the air and guessing, but when that has been the case I've tried to make it clear.

Here are some of the main sources I have used.

Publicly available data sets drawn from process life-cycle analyses

Process-based life-cycle analysis is the most common approach to carbon footprinting. It is often referred to as "bottom-up" because you start off down on your hands and knees, identifying one by one all the processes that have had to happen in order for, say, a product to be created. Then you add up the emissions from each process, and that's the footprint of the product. Simple! Except that it isn't. Not at all. It's back-breaking work, and since the number of processes you really need to count up is always infinite, the job is never quite complete, so you end up with an underestimate. In fact the leaks are often shocking, 50 percent or more. To make matters worse, these problems are popularly overlooked, even in the development of U.K. government-backed and -funded guidelines, such as the PAS 2050 standard (which was published despite a government-commissioned study that concluded that the draft methodology wasn't fit for some of its key intended purposes[3]).

For all the problems, and despite being hard work, process life-cycle analysis is still an essential source of detailed information that can't be gathered any other way. Here are some of the key sources of this type that I've used, each of which is referenced in the main text:

> The U.K. government's Department for Environment, Food and Rural Affairs (Defra) publishes emissions factors for a

range of fuels, electricity sources, transport modes, utilities, and waste. These are mostly U.K. specific and don't take account of full supply chains. I use them where I can but supplement with additions for the missing supply chains. The carbon intensity of electricity in the United States varies from state to state, but it averages out to about 10 percent higher than the U.K. figure, so I have applied that adjustment to all electricity calculations that were originally carried out for the U.K. In Canada the electricity mix is considerably less reliant on fossil fuels than in the U.K. or the U.S., but both for simplicity and because the U.S. and Canada grids are linked, it is spurious to think that turning a light bulb on in Canada has a lower impact than doing so in the United States.

> The University of Bath produces the Inventory of Carbon and Energy, a publicly available data set of carbon emissions factors for hundreds of materials, mainly relating to the construction industry, up to the factory gate.

> The Association of European Plastics Manufacturers (APME) publishes data sets of emissions factors for a wide range of plastics based, not surprisingly, on European manufacture.

> The U.K.'s Market Transformation Programme has a wide range of data on the carbon intensity of common appliances.

> I have drawn on a further wide range of life-cycle analysis studies from all kinds of sources. This is tricky because they all draw their boundaries in slightly different ways and use slightly different assumptions. At its best this has involved me in picking through high-quality academic studies. At its worst it has degenerated into "Google footprinting": scrounging around the web, digging for numbers. When I've sunk to these depths, I've let you know.

Environmental input–output analysis

This is a neat alternative and complement to process life-cycle analysis. It's not as popular, perhaps because it's a bit harder to get your head around, but it's at least as robust as anything else in the murky world of carbon footprinting. It is sometimes called a "top-down" approach because it starts by looking at the whole economy from a height. It uses macroeconomic modeling to understand the way in which the activities of one industry trigger activities and emissions in every other industry. Input-output's key "trick" is a piece of funky math (for which a man called Wassily Leontief got a Nobel Prize) that succeeds in the capturing the endless ripple effects in a way that is 100 percent complete. It has the further advantage that if you know how much you spend on something, you can get an instant crude estimate of its carbon footprint. It's like a magic trick. And just like all the best magic it is also a bit too good to be true: the downside of input-output analysis is that the results can be ridiculously generic.

Input-output analysis is powerful tool both because it doesn't "leak" and because once the model has been built, it is often easy to use. The basic technique is well established. The specific model I've used is one we developed at Small World Consulting with Lancaster University. It draws mainly on data from the U.K.'s Office of National Statistics. Our model is based on a 2007 picture of the U.K. economy; it deals with all the greenhouse gases and employs an emissions weighting factor for high-altitude emissions. The model relies on the key assumption that North American industry has similar carbon intensity per unit of physical output as U.K. industry. This seems reasonable most of the time. A further weakness, which I refer to from time to time and sometimes adjust for, is that it treats imports as though they had the same carbon intensity as domestic production, whereas in reality they are usually more carbon intensive.

Most of the time I have used a combination of process-based and input-output approaches to get my numbers. At their best, process-based methods can be more precise, but input-output analysis is often able to get at places that process life-cycle analysis is unable to reach. Putting the two methods together is sometimes called a hybrid approach, and the result is a bit like looking through both a microscope and a telescope at the same time. They each show you different things, and between them, if the lenses are clean, you might end up with a passable understanding of whatever it is you are looking at.

Booths supermarkets' greenhouse gas footprint model

Over the last three years my company has been mapping out the carbon footprint of the Booths group of U.K. supermarkets and its supply chains. The model we now have draws on a great many life-cycle studies of foods up to the farm gate, often using those funded by Defra. Reports and agricultural models from Cranfield University deserve a mention because I've used them extensively even though they are not uncontentious. Also well worth a mention are five reports produced by the Food Climate Research Network. The Booths model includes transport, processing, packaging refrigeration, and the supermarket chain's other operations. All of these components are attributed to products, broken down into 75 categories. The model goes into a lot of detail, but that doesn't make it accurate. Human understanding of emissions from agriculture is still poor. The model is simply the best picture we have managed to achieve so far. Its purpose is purely practical, and we think it is now good enough to work from, enabling actions to be reasonably well targeted on the hotspots. It is, I think, the most comprehensive model of the climate impacts of supermarket food in the public domain.

Direct greenhouse gas (GHG) emissions
per GDP and per person for 60 countries

Note that these figures do not take account of embodied emissions of imported or exported products, or of international transport. They are simply estimates of the emissions that actually arise from each country.

Country	Population (millions)	GDP (billions of $)	GHG (million tons CO_2e)	GHG per person	GHG/ GDP$	GHG/ GDP as a percentage of U.K. figure	Electricity emission intensity (kg CO_2e/lbs. CO_2e per kilowatt hour)
Argentina	38,730	469	316	0.0082	0.6738	174	0.2750/0.606
Australia	22,121	561	529	0.0239	0.9430	244	0.8680/1.913
Austria	8180	242	91	0.0111	0.3760	97	0.2240/0.493
Belarus	9820	63	74	0.0075	1.1746	304	0.2940/0.648
Belgium	10,420	298	140	0.0134	0.4698	121	0.2740/0.604
Brazil	183,910	1385	983	0.0053	0.7097	183	0.0780/0.171
Bulgaria	7760	58	68	0.0088	1.1724	303	0.4720/1.04
Canada	31,950	919	758	0.0237	0.8248	213	0.2240/0.493
China	1,303,040	7219	6467	0.0050	0.8958	232	0.7710/1.699
Colombia	44,920	300	160	0.0036	0.5333	138	0.1530/0.337
Croatia	4440	50	29	0.0065	0.5800	150	0.3790/0.835
Cyprus	830	17	8	0.0096	0.4706	122	0.8430/1.858
Czech Republic	10,210	182	147	0.0144	0.8077	209	0.5020/1.106
Denmark	5400	159	68	0.0126	0.4277	111	0.3560/0.784
Estonia	1350	18	21	0.0156	1.1667	302	0.7220/1.591
Finland	5230	144	81	0.0155	0.5625	145	0.2970/0.654
France	62,180	1626	563	0.0091	0.3462	90	0.0820/0.180
Germany	82,500	2146	1015	0.0123	0.4730	122	0.4999/1.102
Greece	11,060	226	138	0.0125	0.6106	158	0.7770/1.712
Hungary	10,110	156	83	0.0082	0.5321	138	0.4210/0.928
Iceland	290	9	3	0.0103	0.3333	86	0.0010/0.002
India	1,079,720	3115	1744	0.0016	0.5599	145	0.9120/2.010
Indonesia	217,590	722	470	0.0022	0.6510	168	0.7720/1.701
Iran	67,010	463	583	0.0087	1.2592	326	0.5230/1.153
Ireland	4060	145	68	0.0167	0.4690	121	0.5920/1.305
Italy	58,130	1491	583	0.0100	0.3910	101	0.5240/1.155
Japan	127,690	3435	1355	0.0106	0.3945	102	0.4410/0.972
Kazakhstan	14,990	103	211	0.0141	2.0485	530	1.1160/2.460
Korea (South)	48,080	906	527	0.0110	0.5817	150	0.4370/0.963
Latvia	2310	25	11	0.0048	0.4400	114	0.1810/0.399
Liechtenstein	33	2	–	–	–	0	
Lithuania	3440	41	20	0.0058	0.4878	126	0.1210/0.266
Luxembourg	450	29	11	0.0244	0.3793	98	0.3250/0.716
Malaysia	24,890	235	154	0.0062	0.6553	169	0.4920/1.084

Country	Population (millions)	GDP (billions of $)	GHG (million tons CO_2e)	GHG per person	GHG/ GDP$	GHG/ GDP as a percentage of U.K. figure	Electricity emission intensity (kg CO_2e/lbs. CO_2e per kilowatt hour)
Malta	400	7	3	0.0075	0.4286	111	0.8140/1.794
Mexico	104,000	935	520	0.0050	0.5561	144	0.5760/1.269
Netherlands	16,490	476	218	0.0132	0.4580	118	0.440/0.970
New Zealand	4080	87	75	0.0184	0.8621	223	0.1780/0.392
Nigeria	128,710	137	232	0.0018	1.6934	438	0.4460/0.983
Norway	4590	162	55	0.0120	0.3395	88	0.0090/0.019
Pakistan	152,060	311	230	0.0015	0.7395	191	0.370/0.815
Papua New Guinea	4809	13	7	0.0015	0.5385	139	
Poland	38,180	455	388	0.0102	0.8527	220	0.6620/1.459
Portugal	10,520	189	85	0.0081	0.4497	116	0.4140/0.912
Romania	21,690	169	155	0.0071	0.9172	237	0.4510/0.994
Russian Federation	143,850	1309	1938	0.0135	1.4805	383	0.3290/0.725
Saudi Arabia	23,950	304	371	0.0155	1.2204	316	0.7490/0.651
Slovakia	5380	72	51	0.0095	0.7083	183	0.2550/0.562
Slovenia	2000	38	20	0.0100	0.5263	136	0.3630/0.800
South Africa	45,150	468	505	0.0112	1.0791	279	0.8530/1.880
Spain	42,690	983	428	0.0100	0.4354	113	0.3810/0.839
Sweden	8990	244	70	0.0078	0.2869	74	0.0590/0.130
Switzerland	480	224	53	0.1104	0.2366	61	0.030/0.66
Thailand	63,690	474	320	0.0050	0.6751	175	0.5280/1.164
Turkey	71,790	511	304	0.0042	0.5949	154	0.4960/1.093
U.K.	59,840	1696	656	0.0110	0.3868	100	0.4730/1.042
Ukraine	47,450	279	414	0.0087	1.4839	384	0.3410/0.751
U.S.	293,950	10,708	7065	0.0240	0.6598	171	0.5750/1.267
Venezuela	26,130	145	237	0.0091	1.6345	423	0.2450/0.540
EU (except U.K.) average					0.47053	122	
Non-EU average					0.72497	187	

SOURCE: derived from factsheets within Höhne, N., D. Phylipsen, and S. Moltmann (2007) *Factors Underpinning Future Action: 2007 Update.* A report by Ecofys for the Department for Environment, Food and Rural Affairs. Ecofys GmbH, Cologne. Available at www.fiacc.net/data/fufa2.pdf.

Notes and references

Introduction

1. The phrase "save the planet" is just shorthand for "save the people on the planet." The Earth will be fine until changes in the Sun's radiation evaporate its atmosphere in a billion or so years' time. By this time, as Lord Martin Rees, president of the Royal Society, speculated, the creatures that inhabit the Earth will be as different from people as we are from bacteria.

A quick guide to carbon and carbon footprints

1. *Carbon Footprinting: An Introduction to Organisations,* published by the U.K.'s Carbon Trust (2007) defines on page 1 a carbon footprint in a similar way to me but goes on to describe "basic carbon footprints" on page 4. These are toe-prints rather than rough estimates of footprints.
2. Inventory of U.S. Greenhouse Gas Emissions and Sinks: 1990–2008 (April 2010) U.S. EPA # 430-R-10-006. Available at www.epa.gov/climatechange/emissions/usinventoryreport.html
   The U.K.'s 2009 Environmental Accounts give a similar breakdown: carbon dioxide (86 percent), methane (7 percent), nitrous oxide (6 percent), and refrigerant gases (1 percent).
3. All the gases covered by the Kyoto Protocol are included. For better or worse, I have adopted a common convention of considering the impact over a 100-year period. This makes a difference because some gases last longer than others. CO_2 stays in the atmosphere for a very long time, whereas methane and refrigerant gases fade much faster. If we were measuring the impact over 20 years, methane would be about three times more prominent.
4. There is much scientific uncertainty around the impact of high-altitude emissions. A figure of 1.9 can be inferred from the IPCC 4 assessment report. This is also the figure suggested in the 2009 Guidelines to Defra/DECC's Greenhouse Gas Conversion Factors for Company Reporting (Annex 6, footnote 10).

5. Here is some more detail on US and Canadian targets: Internationally, America has committed to cut GHG emissions to 4 percent below 1990 levels by 2020 under the Copenhagen Accord; however, this commitment has to be backed be legislation, which has not yet passed the Senate. See www.scientificamerican.com/article. cfm?id=us-commits-to-greenhouse-gas-cuts-under-copenhagen-accord.

Senators Kerry and Lieberman released their draft discussion of their American Power Act on 12 May 2010, a bill which "establishes goals for economy-wide emission reductions from 2005 levels: 4.75 percent by 2013, 17 percent by 2020, 42 percent by 2030, and 83 percent by 2050." See www.pewclimate.org/federal/analysis/congress/111/ short-summary-american-power-act-kerry-lieberman.

Nationally, President Obama has signed an Executive Order on Federal Sustainability, committing the Federal Government to lead by example and reduce greenhouse gas emissions by 28 percent by 2020 from 2008 levels, increase energy efficiency, and reduce fleet petroleum consumption. (See www.whitehouse.gov/issues/energy-and-environment, www.whitehouse.gov/the_press_office/President-Obama-signs-an-Executive-Order-Focused-on-Federal-Leadership-in-Environmental-Energy-and-Economic-Performance, and www.whitehouse.gov/assets/ documents/2009fedleader_eo_rel.pdf.)

"The Government of Canada is committed to reducing Canada's total greenhouse gas emissions by 17 per cent from 2005 levels by 2020. This target is completely aligned with the U.S. target, and is subject to adjustment to remain consistent with the U.S. target." (See www.ecoaction. gc.ca/climatechange-changementsclimatiques/index-eng.cfm.)

6. How many tons for a life? Here is a back-of-the-envelope calculation.

I'm not presenting this as even the beginnings of a rigorous argument but just as a line of thought that might serve to make that critical connection between carbon and life itself.

Let's see whether it's possible to estimate of the number of tons of carbon released into the sky before it would be reasonable to think that someone somewhere is going to have to die as a consequence.

It will be simplistic, but I hope you will stick with it. I'm going to take a high-emissions scenario and a low-emissions scenario for the world and make assumptions about how many people will die as a result of climate change from each. Then I'll take the difference in the carbon emissions, divide it by the difference in the number of deaths around the world and use that as the guide to the number of tons per death.

For my low-emissions scenario, I'm going to assume that we immediately cut global emissions by 40 percent to 30 gigatons CO_2e—that's 30 billion tons—and hold them like that for 40 years. (Clearly that is impossible. I'm going to use this scenario because the sums are simple. A more

realistic scenario with a similar climate outcome might be a rapid reduction starting today and resulting in global emissions falling to about 10
gigatons by 2050. This would still be an extremely radical response and
one that the world hardly seems on the brink of adopting.)

Let's assume that under the low-emissions scenario relatively few
people—a small fraction of a billion—die as a result of climate change. That
looks likely, although there is still a risk that the outcome would be worse.
One estimate is that the death rate related to climate change is already
300,000 per year. We know that the climate is going to continue to get
warmer for years even if we cut our emissions to zero right now, because
of the greenhouse gases that are already in the atmosphere, so that
300,000 figure is unlikely to decrease. Still, it seems entirely plausible to
keep things down to such a small fraction of a billion that for the purpose
of this estimate we can call it zero.

For my high-emissions scenario I'm going to assume that until 2050
we average the current 50 gigatons CO_2e per year. Many scientists think
our species will be in a lot of trouble if that happens. It's not unrealistic
to think that we might average the current 50 gigatons when you think
that right now the trend in global emissions is not just rising fast but still
accelerating (tempered only by the short-term blip of the global reces-
sion). One view is that in that kind of scenario, by 2100 the world will
only be able to support 1 billion people instead of the 9 billion of us that
are forecast without climate change.

(To visualize the time lags involved in reversing climate change, imagine someone in a car, accelerating like crazy and already way above the
speed limit when they notice that they are in an "average speed check
area." To avoid the fine, the driver has to first switch his foot from accelerator to brake, then wait for the car to slow down and then drive slowly
for quite a while until their average speed slips below the limit. Each step
in the process takes time. It's the same for the world, only far worse. The
rate of emissions corresponds to the power going into the engine, but it
takes years for the global community even to start wondering whether
to begin the process of adjusting the position of its feet on the pedals.
Once that has been decided and the foot has been shifted, the long and
difficult process of braking can begin. Even when, many years later, the
stuff coming out of our chimneys, fields, and forest fires around the world
has dropped to a low enough level, the temperature will carry on rising
until the total amount of carbon in the reservoir of the sky has dropped
by enough for the temperature to begin to fall. Even then any remaining ice will carry on melting until the temperature has actually dropped.
Whatever we do now, falling temperatures are decades away, and no one
reading this book is likely to see increasing ice. And right now we are still
a long way from having our foot on the right pedal. I've over-simplified
the science, but the basic concept is right.)

So that is a difference of 8 billion people. (The Optimum Population Trust estimates a mere 5 billion difference, but their figures are in the same ball park as mine: "Earth heading for 5 billion overpopulation?" [Optimum Population Trust, March 2009], www.optimumpopulation. org/releases/opt.release16Mar09.htm.) It is not fair to say that all those 8 billion will have to die as a result of climate change, because some will simply never be born, so let's say that in this scenario, 4 billion people will die. It's chilling, and you may not buy into the argument completely. None of us knows exactly what would happen, and we can't actually run both scenarios to find out. But I'm going to run with these numbers for the sake of the thought experiment.

The difference between the scenarios is 600 billion tons of CO_2e and 4 billion deaths. That works out to one death per 150 tons CO_2e.

7. £12 ($18) per ton CO_2e is the maximum price that a company might have to pay under the U.K.'s Carbon Reduction Commitment. The European Trading Scheme puts it at about half of that.

Under 10 grams

1. Drawn from a Swedish life-cycle assessment study in 2004 by Mireille Faist Emmenegger, Rolf Frischknecht, Markus Stutz, Michael Guggisberg, Res Witschi, and Tim Otto: "Life cycle assessment of the mobile communication system UMTS: toward eco-efficient systems," www.esu-services. ch/download/faist-2005-UMTS.pdf.

2. Gartner press release, 2008: "Gartner says mobile messages to surpass 2 trillion messages in major markets in 2008," www.gartner.com/ it/page.jsp?id=565124. With 1.9 trillion messages in 2007, they predicted 2.3 trillion in 2009. I have extrapolated a bit further.

3. The American Water Works Association, gives a figure of 69.3 gallons per capita per year for indoor per capita water use with the following breakdown:

Use	Gallons per capita	Percentage of total daily use
Showers	11.6	16.8
Clothes washers	15.0	21.7
Dishwashers	1.0	1.4
Toilets	18.5	26.7
Baths	1.2	1.7
Leaks	9.5	13.7
Faucets	10.9	15.7
Other domestic uses	1.6	2.2

SOURCE: www.drinktap.org/consumerdnn/Home/WaterInformation/ Conservation/WaterUseStatistics/tabid/85/Default.aspx (accessed July 2010).

The average in the U.K. is about 150 liters (40 gallons) per person per day, according to Defra's domestic water consumption summary at www.defra.gov.uk/sustainable/government/progress/regional/summaries/16.htm.

The carbon from driving is based on 18.6 miles per gallon and includes fuel supply chains as well as the embedded carbon in the vehicle itself (See Driving 1 mile, page 66)

4. "The energy needed to treat and pump mains water to our homes, and to collect and treat waste water from the sewage network, is responsible for nearly 1 percent of the U.K.'s annual greenhouse gas emissions." U.K.'s Energy Saving Trust, "Water and carbon—the facts," www.energysavingtrust.org.uk/Water/Water-and-carbon-the-facts. EST's figure for total U.K. emissions excludes the overseas component of the footprint. Small World Consulting's input–output model puts the greenhouse gas emissions from household water supply at 0.3 percent and emissions from sewage and sanitary services at nearly 2 percent, but that includes other things. A detailed explanation of the methodology is contained in: M. Berners-Lee, D.C. Howard, J. Moss, K. Kaivanto, and W.A. Scott, "Greenhouse gas footprinting for small businesses—the use of input-output data," 2011, *Science of the Total Environment.*

5. Based on figures for the carbon intensity of U.K. water supply and treatment: Defra (2009), "Guidelines to Defra's GHG conversion factors for company reporting," www.defra.gov.uk/environment/business/reporting/pdf/20090928-guidelines-ghg-conversion-factors.pdf (accessed February 2010).

6. Word for word, the Google story, from the company's blog, is: "Queries vary in degree of difficulty, but for the average query, the servers it touches each work on it for just a few thousandths of a second. Together with other work performed before your search even starts (such as building the search index), this amounts to 0.0003 kWh [kilowatt-hours] of energy per search, or 1 kJ [kilojoule]. For comparison, the average adult needs about 8,000 kJ a day of energy from food, so a Google search uses just about the same amount of energy that your body burns in ten seconds. In terms of greenhouse gases, one Google search is equivalent to about 0.2 grams of CO_2." From googleblog.blogspot.com/2009/01/powering-google-search.html.

7. This estimate is based on a breakdown of power usage at Lancaster University.

8. For example, Dr. Alex Wissner-Gross, a physicist from Harvard University, wrote in *The Sunday Times* on January 9, 2009: "How you can help reduce the footprint of the Web," www.timesonline.co.uk/tol/news/environment/article5488934.ece.

9. The sums: 5 m^2 (54 sq. ft.) doorway, fully open for 15 seconds, wind speed through the door of 1 meter (3 feet) per second, temperature difference of 15°C (59°F), heat capacity of air 1.2 kilojoules per cubic meter

(35 joules per cubic foot), heat supplied by gas at 0.22 kg (0.48 lbs.) CO_2e per kilowatt-hour.

10. BREEAM: Building Research Establishment Environmental Assessment Method. I understand that the BRE has since improved its energy efficiency criteria somewhat. BREEAM is, roughly speaking, a U.K. equivalent of LEED (Leadership in Energy and Design) although arguably more academic in flavor and more complex. The sums here are based on a temperature difference of 15°C (59°F)—typical for winter—and a wind speed of just 2.5 miles per hour flushing warm air out of the building.

11. I am assuming that this low-grade paper comes in at just 1 kg CO_2e per kilo (1 lb. CO_2e per pound).

12. Association of Plastic Manufacturers. Eco-profiles showing emissions from production of a wide variety of plastics are available from lca. plasticseurope.org/index.htm (accessed 20 April 2008). Based on 3 g weight per bag.

13. See www.reusablebags.com/facts.php. Vincent Cobb's website (www.reusablebags.com) contains interesting data on the numbers of bags used around the world, as well as their impacts.

10 grams to 100 grams

1. Annex 9 in Defra (2009), "Guidelines to Defra's GHG conversion factors for company reporting" (www.defra.gov.uk/environment/business/ reporting/pdf/20090928-guidelines-ghg-conversion-factors.pdf), gives a figure of 550 kg (1,210 lbs.) CO_2e per ton of paper and cardboard in landfill.

2. www.caloriesperhour.com.

3. In our input-output model of the greenhouse gas footprint of U.K. industries (for the chapter entitled "Under 10 grams"), sports goods typically have a carbon intensity of around 250 g per pound's worth of goods at retail prices. If we make the very broad assumption that cycling goods are typical of this, and if we say that Her Majesty's Revenue and Customs (HMRC) is being roughly fair to reimburse you at 20p (31 cents) per mile for business travel on a bike, then we would need to add about 50 g CO_2e per mile to take account of the wear and tear on your bike, your waterproof gear, lights, helmet, and so on. Actually, as someone one who is frequently cycling between offices and train stations trying to keep jacket, tie and laptop dry, I suspect that HMRC has underestimated it and should be paying out the full 40p (62 cents) per mile that they allow for car users. (This would also provide a beneficial incentive.)

4. C. Saunders, A. Barber, and G. Taylor (2006), *Food Miles— Comparative Energy/Emissions Performance of New Zealand's Agriculture Industry.* Research Report no. 285. Lincoln, New Zealand: Lincoln University.

5. Blanke and Burdick (2005), referenced in Defra (2006), *Environmental Impacts of Food Production and Consumption,* p. 47, randd.defra.gov.uk/ Document.aspx?Document=EV02007_4601_FRP.pdf.

6. For the footprint up to the farm gate I've used a number from A. Wallen, N. Brant, and R. Wennersten (2004), "Does the Swedish consumer's choice of food influence greenhouse gas emissions?" *Envir Sci Policy* 7, 525–535. For the rest of the footprint I've used my work at Booths Supermarkets. The main report is available online: "The Greenhouse Gas Footprint of Booths, July 2010," www.booths.co.uk/Documents/Booths_ Full_Report_100720.pdf.

7. I bought a bunch in a plastic bag weighing just 4 g, the reason for which is to stop customers from ruining them when they try to split a bunch. So the bag could be worth it until we all learn not to maul the fruit.

8. The Oxfam Cool Planet website has a simple (and child friendly) account of how fair-trade bananas are grown in the Windlass Islands, along with recipes: www.oxfam.org.uk/coolplanet/kidsweb/banana/ index.htm.

9. There is more on this at the very accessible Banana Link website "Working toward a fair and sustainable banana trade," www.bananalink. org.uk. For a critical and pessimistic look at the future of bananas in our lives see also Dan Koeppel (2008) "Yes, we will have no bananas," *New York Times* (18 June), www.nytimes.com/2008/06/18/opinion/18koeppel. html?_r=1. Waitrose has commissioned a life-cycle analysis of one of its banana supply chains through the University of Bangor, due for publication in late 2010, I gather. I understand that this has taken account of deforestation issues and will therefore make an interesting read when it becomes available.

10. Food Production Daily, November 2004, www.foodproduction- daily.com/Supply-Chain/Half-of-us-food-goes-to-waste.
   The same source quotes U.S. consumers wasting just 14 percent. This feels conservative given recent research by WRAP in estimating the figure at over 20 percent for the U.K.

11. My numbers for the footprint up to the farm gate come from two sources: one of oranges grown in Spain, the other of produce for the Swedish market: N. Sanjuan, L. Ubeda, G. Clemente, and A. Mulet (2005), "LCA of integrated orange production in the Comunidad Valenciana" (Spain). *Int J Agric Resources Governance Ecol* 4 (2), 163–177; and A. Wallen, N. Brant, and R. Wennersten (2004), "Does the Swedish consumer's choice of food influence greenhouse gas emissions?" *Envir Sci Policy* 7, 525–535. For the rest of the footprint I've used our work at Booths.

12. U.S. TV-WATCHING STATISTICS:
   > According to the A.C. Nielsen Co., the average American watches more than 4 hours of TV each day (or 28 hours/week, or 2 months of nonstop TV-watching per year). In a 65-year life, that person will have spent 9 years glued to the tube.
   > Number of TV sets in the average U.S. household: 2.24
   > Percentage of U.S. homes with three or more TV sets: 66

> Number of hours per day that TV is on in an average U.S. home: 6 hours, 47 minutes
   Source: www.csun.edu/~vceed002/health/docs/tv&health.html
   CANADIAN TV-WATCHING STATISTICS:
> Canadians watch an average of 21.4 hours of TV a week (population 2 years old and over)
   Source: www40.statcan.ca/l01/cst01/arts23-eng.htm
   EUROPEAN TV-WATCHING STATISTICS:
   December 2009: "The U.K. witnessed the highest average increase in TV watching during 2008, up by 3.2 percent to 3.8 hours a day. This was higher than the average (3.5 hours per day) across the European countries surveyed, but still slightly less than viewers in Italy, Poland, and Spain. U.S. viewers consumed the most television in 2008, watching on average 4.6 hours a day, up 1.8 percent from 2007, while viewers in Sweden continued to watch the least at 2.7 hours a day, although this was a 1.9 percent increase across the year."
   Source: www.ofcom.org.uk/media/news/2009/12/nr_20091217.
13. All the data on TV power consumption come from Ireland's Electricity Supply Board (2009), www.esb.ie/main/sustainability/energy-services.jsp. I have allowed 10 g per hour for the satellite receiver.

100 grams to 1 kilo (2.2 pounds)

1. The London bus occupancy factor is from the Transport for London 2008 environment report available at www.tfl.gov.uk/assets/downloads/corporate/environment-report-2008.pdf and www.tfl.gov.uk/assets/downloads/corporate/environment-report-2008-data-tables.pdf.
2. S. Aumônier, M. Collins, and P. Garrett (2008), *An Updated Life-cycle Assessment Study for Disposable and Reusable Nappies*, Science Report no. SC010018/SR2. U.K. Environment Agency. randd.defra.gov.uk/Document.aspx?Document=WR0705_7589_FRP.pdf (accessed 1 March 2010).
3. Data from T. Garnett (2006), *Fruit and Vegetables & U.K. Greenhouse Gas Emissions: Exploring the Relationship.* Food Climate Research Network, Surrey; and C. Foster, K. Green, M. Bleda, P. Dewick, B. Evans, A. Flynn, and J. Mylan (2006), *Environmental Impacts of Food Production and Consumption: A Report to the Department for Environment, Food and Rural Affairs.* Manchester Business School. Defra, London.
4. Direct emissions from fuel and electricity generation and supply come from Defra (2009) *Guidelines to Defra/DECC's GHG Conversion Factors for Company Reporting.* Produced by AEA for the Department of Energy and Climate Change (DECC) and the Department for Environment, Food and Rural Affairs (Defra). Supply chains and infrastructure are estimates made from the input–output model. (See above note 4 in the "Under 10 grams" section.)

5. David J. C. MacKay lays out the math nicely in *Sustainable Energy without the Hot Air* (2009), published by UIT Cambridge Ltd and available as a free download on www.withouthotair.com.

6. R. Kemp (2007), *Traction Energy Metrics*. Rail Safety & Standards Board, London. Available from www.rssb.co.uk/pdf/reports/research/T618_traction-energy-metrics_final.pdf.

7. The carbon intensity of PET, from which bottles are typically made, is 3.3 kg CO_2e per kilo, based on figures from the Association of Plastic Manufacturers, *Eco-profiles of the European Plastics Industry—Main Flow Chart*, available from lca.plasticseurope.org/index.htm (accessed 20 April 2008). The bottles I have weighed average around 50 g per quart of capacity. The transport footprint is based on a U.K. typical rigid lorry, and conversion factors also from DEFRA—and is therefore light by up to 50 percent because fuel supply chains and the embodied footprint of the vehicle are not taken into account. I took the energy of bottle manufacture to be around one-quarter of the energy required to make the PET pellets, on the basis of Stefano Botto, "Tap water vs. bottled water in a footprint integrated approach" (July 2009), precedings.nature.com/documents/3407/version/1.

8. The consumption figures came from Andrea Thompson, "The energy footprint of bottled water," *Live Science*, 19 March 2009, www.livescience.com/environment/090318-bottled-water-energy.html (accessed October 2009). According to this, most is consumed in the U.S. and Europe, with the U.S. accounting for 8.7 billion gallons (29 gallons per person).

9. Confederation of Paper Industries (2006), *U.K. Paper Making Industries Statistical Facts Sheet;* Utrecht Centre for Energy Research (2001) ICARUS-4: *Sector Study for the Paper and Board Industry and the Graphical Industry,* available from copernicus.geog.uu.nl/uce-uu/downloads/Icarus/Paper.pdf (accessed 3 April 2008).

10. Defra (2009), "Guidelines to Defra's GHG conversion factors for company reporting" (www.defra.gov.uk/environment/business/reporting/pdf/20090928-guidelines-ghg-conversion-factors.pdf), Annex 9.

11. Shelly Shumacher in ehow.com (www.ehow.com/how_5072539_stop-usps-junk-mail.html). Shumacher writes:

> There is no one service for the eradication of junk mail, but there are several things you can do to reduce the amount received.
>
> Black out the bar code and address on all junk mail that is delivered using first class postage. Put a circle around the postage and write "Not accepted: return to sender." This can be put in any mail box, and it will be returned to the business that sent it.
>
> Request a 1500 form from the U.S. Postal Service to stop the delivery of sexually explicit material.

Call 1-800-5 OPT OUT to stop mail generated from the three major credit reporting agencies. This will stop the majority of credit card offers.

Contact by phone, mail or email all companies that send out catalogs and request to be taken off their mailing lists.

Call 1-800-645-9242 to be taken off the Publisher's Clearinghouse Sweepstakes Mailing list.

Contact American Family at 1-800-237-2400 to be removed from its sweepstakes mailing list.

Call or write the Direct Marketing Association (see Resources) and ask them to activate the preference service. This will eliminate close to three quarters of all direct mail for 5 years. You will need to sign up again after 5 years.

12. The Red Dot Campaign (www.reddotcampaign.ca/) is working to reduce the amount of junk mail sent out. They list some ways to reduce the amount of junk mail sent:

Tape a "No Junk Mail" sign on or in your mailbox or mail slot.

OPTIONAL: If your sign is ignored, download, sign, and date a "no junk mail" letter from www.reddotcampaign.ca/downloads/RedDot-Letter-toCanadaPost.pdf and mail to your nearest Canada Post outlet. For more information about its Consumer's Choice program, contact Canada Post or call 1-866-607-6301.

Sign up with the Canadian Marketing Association's "Do Not Contact" registry. This enables you to reduce the number of marketing offers received by mail, telephone, and fax.

Sign up for Canada Post's e-post at www.canadapost.ca/cpo/mc/business/campaigns/epost/default.jsf?LOCALE=en and help save even more paper by getting your bills online.

13. All the numbers on waste impacts come from Defra Annex 9 (see note 10 above), released in September 2009, and are based on the U.K. Conversion factors for virgin and recycled paper came from Confederation of Paper Industries (2006) (see note 9 above). Environmental Defence Fund (1995) "Energy, air emissions, solid waste outputs, waterborne wastes and water use associated with component activities of three methods for managing newsprint" provided a sense and some figures for transport and printing impacts.

14. Based on U.K. electricity at 0.6 kg (1.3 lbs) CO_2e and gas at 0.225 kg (0.49 lbs) CO_2e per kilowatt-hour. Both figures are based on those supplied by Defra (2009) but are adjusted to take account of power station supply chains and distribution. The cost of electricity is taken as 10p per kilowatt-hour.

15. Figures from Defra's (U.K.) Boiler Efficiency Database, 2010 (www.sedbuk.com) are shown in the table below. £1 is around $1.5.

Furnace type	Seasonal efficiency (percent)	TYPICAL ANNUAL FUEL COST (£)				
		Flat	Bungalow	Terraced	Semi-det'd	Detached
Old (heavy weight)	55	£267	£341	£354	£397	£550
Old (light weight)	65	£231	£293	£304	£340	£470
New (non-condensing)	78	£197	£249	£258	£289	£396
New (condensing)	88	£178	£224	£232	£259	£355

16. Based on Defra (2008), "Guidelines to Defra's GHG conversion factors for company reporting." Available from www.defra.gov.uk/environment/ business/reporting/pdf/ghg-cf-guidelines-annexes2008.pdf, supplemented with estimates derived from an environmental input–output model of emissions from supply chains (see above note 4 in the "Under 10 grams" section) that are not included in the Defra figures. These additions account for around 10 percent of the total footprint of U.K. electricity.
17. Environment Canada, Electricity Intensity Tables. I've added 10 percent to the national average to take account of supply chains. If you live in a less carbon-intense province, remember that provinces trade with each other and the marginal demand is met through fossil fuel.
18. Public opinion research by Landor Associates (2007), *ImagePower Green Brands Survey*, available from www.landor.com/?do=news. pressrelease&r=&storyid=508, found that whereas "the adoption of green values is the fastest consumer trend in recent years, faster than the uptake of the Internet or mobile phone," the public still lacks a sophisticated understanding of what it actually means to be green. The color of the logo was frequently taken as one of the key indicators of green credentials.
19. This figure is drawn from my work with Booths supermarkets. The main report is available online: "The Greenhouse Gas Footprint of Booths, July 2010," www.booths.co.uk/Documents/Booths_Full_Report_100720.pdf.
20. This is an average figure. Bargain flights can be a lot worse. If you get your Chicago to Dallas (800 miles) return for $10, that's around 50 kg (110 lbs.) CO_2e per dollar. Then again, if you pay over, odds are that it's less carbon intensive.
21. I've used a figure of 0.68 kg (1.5 lbs.) CO_2e per mile for an average gas-powered car (including fuel supply chains and manufacture of the car). Exhaust pipe emissions from Defra (2009), *Guidelines to Defra/DECC's GHG Conversion Factors for Company Reporting. Produced by AEA for DECC and Defra*. Available from www.defra.gov.uk/environment/business/reporting/pdf/20090928-guidelines-ghg-conversion-factors.pdf.
22. Tim Jackson (2009), *Prosperity without Growth: Economics for a Finite Planet*. Earthscan, London.
23. To keep things simple, I'm not going to look at the 12 percent of general waste that gets incinerated instead of being sent to landfill. It doesn't change the overall picture much.

24. The following data is from U.S. Environmental Protection
    Agency (available at www.epa.gov/epawaste/nonhaz/municipal/pubs/
    msw2008rpt.pdf)
       Recycling (also called recovery) and disposal
       % of trash recycled and composted in 2008: 33.2%
       Composted waste in 2008: 22.1 million tons
       Recycled waste (not incl. composted) in 2008: 61 million tons
       Total municipal solid waste in 2008: 250 million tons
       I've taken the U.S. population to be 293 million and assumed that the
    carbon impact per ton sent to landfill is the same as in the U.K.
25. All the numbers of waste impacts come from Defra Annex 9 (see
    note 10 above), and are based on the U.K. data on U.K. consumer waste
    from the Office of National Statistics website, www.statistics.gov.uk/cci/
    nugget.asp?id=1769 (accessed October 2009). The breakdown of con-
    sumer waste by type comes from the BBC (July 2007), "Household waste:
    in statistics," news.bbc.co.uk/1/hi/uk/6222288.stm.
26. U.K. Market Transformation Programme: *BNW16: A Comparison
    of Manual Washing Up with a Domestic Dishwasher*, www.mtprog.com/
    spm/download/document/id/598 (accessed October 2009).
27. I based this on a machine costing $525, lasting 10 years, and used
    120 times per year, and a figure of 0.66 kg (1.46 lbs.) CO_2e per dollar
    expenditure on domestic appliances at 2008 retail prices from the input-
    output model. (See above note 4 in the "Under 10 grams" section.)
28. The U.K.'s Market Transformation Programme study (see note 26
    above) points to a possible carbon savings of about one-fortieth from a
    decrease in water use. A whole bottle of dishwashing detergent probably
    has a footprint of about 1 kg (2.2 lbs.) CO_2e.
29. www.toiletpaperworld.com.
30. Worldwatch Institute (2007), *The Reality behind Toilet Paper Con-
    sumption*, www.worldwatch.org/node/5162 (accessed October 2009).
31. Tesco press release, 1 May 2009: "Tesco carbon labels toilet
    paper," www.tescoplc.com/plc/corporate_responsibility_09/news/
    press_releases/pr2009-05-01/.
32. The figures come from models used by Small World Consulting
    (www.sw-consulting.co.uk). An input-output approach (see above
    note 4 in the "Under 10 grams" section) is used for the fuel supply
    chains and the depreciation of the embodied emissions in the car and
    its manufacture.
33. Derived from Defra (2008), *Passenger Transport Emissions
    Factors: Methodology Paper*. Available from www.defra.gov.uk/
    environment/business/reporting/pdf/passenger-transport.pdf
    (accessed 3 April 2008).
34. The (slightly simplified) physics of the energy per mile required
    to move a car looks like this: Energy per mile = Energy required to over-
    come rolling resistance + Energy required to overcome air resistance.

The rolling resistance component is independent of your speed, but the energy per mile required to overcome the air resistance goes up with the square of your speed. At highway speeds the rolling resistance fades out of the picture in comparison with the air resistance, so the total energy per mile, and hence the fuel consumption, becomes proportional to the square of the speed. Therefore dropping your speed by one-seventh gets you an improvement in miles per gallon of more than a quarter. See MacKay (2009) (note 5 above).

35. A. Williams (2007), "Comparative study of cut roses for the British market produced in Kenya and the Netherlands. Précis report for World Flowers, 12 February 2007," www.fairflowers.de/fileadmin/flp. de/Redaktion/Dokumente/Studien/Comparative_Study_of_Cut_Roses_ Feb_2007.pdf. The numbers for Kenyan and Dutch roses were derived from this study by Cranfield University. Note that the study was commissioned by World Flowers, which imports lots of flowers from Kenya. Note also that only the précis report was made public. We adjusted the figure for Kenyan roses down slightly because Cranfield had applied an emissions weighting factor of 2.7, in contrast with the 1.9 that we have been using in this book. However, the report uses generic air freight conversion factors supplied by Defra. A final note of caution is that the generic air freight conversion factors used do not take account of inefficiencies due to the bulkiness of flowers. Despite all these reservations, I think the broad conclusion of the report is probably right, and on the back of it, Booths is sourcing its Valentine's Day roses from Kenya instead of Holland this year.

36. Emissions up to the farm gate are from A.G. Williams, E. Audsley, and D.L. Sandars (2006), *Determining the Environmental Burdens and Resource Use in the Production of Agricultural and Horticultural Commodities.* Main Report, Defra Research Project IS0205. Cranfield University, Bedford, and Defra. Available on www.silsoe.cranfield.ac.uk and www.defra.gov.uk. Beyond the farm gate, based on study by Small World Consulting for Booths Supermarkets. The main report is available online: "The Greenhouse Gas Footprint of Booths, July 2010," www.booths.co.uk/ Documents/Booths_Full_Report_100720.pdf.

37. International Energy Agency (2007), *Tracking Industrial Energy Efficiency and CO_2 Emissions,* www.iea.org/work/2007/tracking/ conference_proceedings.pdf. Small World Consulting's input–output model (see above note 4 in the "Under 10 grams" section) suggests that direct emissions and electricity between them account for just 91 percent of the footprint of the U.K. cement industry. I have therefore added 10 percent to the IEA figures to take account of energy supply chains and other indirect emissions in the cement industry.

1 kilo to 10 kilos (2.2 pounds to 22 pounds)
1. I'm using conversion factors of 1.59, 2.59, and 2.70 kg CO_2e per kilo (1.59, 2.59, and 2.70 lbs. CO_2e per pound) for printing on recycled, typical U.K.

mix and 100 percent virgin paper. Confederation of Paper Industries (2006) U.K. Paper Making Industries Statistical Facts Sheet, www.paper.org.uk/info/reports/fact2006colour0707.pdf (accessed 3 April 2008).

2. It's difficult to talk about the footprint of a new product innovation, because it all depends on how much of the R&D and tooling up you assign to the first models. They currently sell at around $225, but those prices will probably tumble. The input–output model (see above note 4 in the "Under 10 grams" section) gives us a carbon intensity of around 0.4 kg (0.9 lbs.) CO_2e per U.S. dollar for the computer industry. That's a footprint of 90 kg (198 lbs) CO_2e, but a lower figure of 50 kg (110 lbs) is probably more realistic, reflecting the impact of mass production.

3. From WRAP (2008), *The Food We Waste*. Waste & Resources Action Programme (WRAP), Banbury. Available at www.wrap.org.uk/down-loads/The_Food_We_Waste_v2_2_.ec417f6c.5635.pdf. The overall figure of 30 percent waste includes bones and other bits that we don't all consider edible.

4. All the numbers here come from the Association of Plastic Manufacturers, Eco-profiles of the European Plastics Industry, available from lca.plasticseurope.org/index.htm (accessed 20 April 2008).

5. 2K Manufacturing is scheduled to start bulk production in March. I am typing amid the 2009 pre-Christmas snow—another example of the strange time warp that exists between us.

6. A typical bath can hold about 32 gallons (with you in it too). I've taken the cold water temperature to be 8°C (46°F) and a comfortable bath temperature to be 39°C (102°F). The heat capacity of water is 4.2 kilojoules per quart. I've assumed a 90 percent efficient furnace and used a conversion factor of 0.225 kg (0.50 lbs.) CO_2e per kilowatt-hour for heat produced by natural gas (this uses a figure from Defra for the direct emissions of burning gas and adds to that a figure from our input–output model to estimate the supply-chain impacts). There are 3,600 kilojoules per kilowatt-hour. The footprint of the bath in kg CO_2e is $120 \times (39-8) \times 4.2 \times 0.225 / (3,600 \times 90$ percent$)$. The footprint of the water consumption is negligible.

7. At the time of writing, I'm told that Booths is switching its small green beans from Peru to nearer-by Egypt. It's still air freight but a big reduction nonetheless.

8. Input–output modeling (see above note 4 in the "Under 10 grams" section) gives a guideline of 440 g CO_2e per pound sterling's worth of expenditure on domestic appliances (293 g per U.S. dollar). I have assumed that each is good for 1,000 uses and based my calculation on washing machines costing £300 ($450), and tumble driers £200 ($300).

9. The rating system seems a bit unfair on condensing driers, since it doesn't take account of the fact that they keep the heat in the home instead of belching it into the outside world.

10. Based on 2,500 calories for a man and 2,000 calories for a woman.

11. A.G. Williams, E. Audsley, and D.L. Sandars (2006), *Determining the Environmental Burdens and Resource Use in the Production of Agricultural and Horticultural Commodities*. Main Report. Defra Research project IS0205. Cranfield University, Bedford, and Defra. Available from www.silsoe.cranfield.ac.uk and www.defra.gov.uk.
12. Best guess based on the "average figure," which I have derived simply as a round number midway between efficient and inefficient production.
13. International Rice Research Institute, beta.irri.org.
14. International Rice Research Institute, beta.irri.org. See also Fertilizer, page 136.
15. Rice statistics are shown in the table below. Sources: International Rice Research Institute, beta.irri.org; Worldwatch Institute (2009), *State of the World 2009: Confronting Climate Change*, 26th ed., Earthscan, London. Total consumption figure taken from 2008, total fertilizer figure from 2005.

CARBON FOOTPRINT OF RICE PRODUCTION	Low estimate	High estimate
Global rice consumption (million tons)	432	432
Fertilizer applied (million tons)	161	161
Percentage of global calories provided	19.7	19.7
Annual methane from paddy fields (million tons CO_2e)	600	600
Emissions per kilo (2.2 pounds) of fertilizer (kg CO_2e)	2.7	12.3
Emissions from fertilizer use (million tons CO_2e)	435	1,984
Transport and other agricultural emissions (million tons CO_2e)	43	43
Footprint per kilo of rice (kg CO_2e)	2.5	6.1
Footprint per pound of rice (kg CO_2e)	1.1	2.8
Agricultural footprint if all the world's calories came from rice (million tons CO_2e)	5,476	13,340

16. Data taken or derived from WWF (2007), *Making Water: Desalination: Option or Distraction for a Thirsty World?* Available at assets.panda.org/downloads/desalinationreportjune2007.pdf (accessed October 2009).
17. According to a report by the European Environment Agency, Spain and Portugal will be most affected within the EU by the coming climate change. Storms, floods, and droughts are likely to become more and more frequent. J.M. Moreno, et al. (2005), *A Preliminary General Assessment of the Impacts in Spain Due to the Effects of Climate Change*. Ministerio de Medio Ambiente. Available at www.mma.es/secciones/cambio_climatico/areas_tematicas/impactos_cc/pdf/evaluacion_preliminar_impactos_completo_2.pdf.

18. Information on Seawater Greenhouse from Wikipedia, en.wikipedia. org/wiki/Seawater_greenhouse. My calculations also used figures on U.K. water consumption from the U.K. government: "Indicators of sustainable development," www.sustainable-development.gov.uk/sustainable/quality04/maind/04q02.htm (accessed October 2009), and a figure of 0.6 kg (1.3 lbs.) CO_2e per kilowatt-hour of electricity to drive pumps. None of my analysis includes any of the electrical energy that is required only to pump water to and from the sea. Seawater Greenhouse figures come out at 3.9 kg (8.6 lbs) CO_2e per ton, on the basis of a 300-mile journey with a 100-meter (330-foot) height gain, using grid electricity from a fuel mix similar to that of the U.K.'s and not taking into account the possibility for recovering some of the energy from the brine returning downhill through turbines.

19. E.M. Kalliala and P. Nousiainen (1999), "Life cycle assessment: environmental profile of cotton and polyester-cotton fabrics." AUTEX Res J 1(1), 8–20.

20. Based on a study on cotton towels, which found that over the two-year lifetime of the towel, the laundry was going to have three and a half times the impact of the creation of the towel. I've assumed that my jeans will last longer, but I'm nevertheless suspicious that the embodied emissions have been underestimated in the study. I've assumed that these two will cancel each other out. R. Blackburn and J. Payne (2004), "Life cycle analysis of cotton towels: impact of domestic laundering and recommendation for extending periods between washing." Green Chem 6, G59–G61.

21. Association of Plastic Manufacturers (see note 4 above).

22. A report by the Stockholm Environment Institute estimated that it takes between 9,788 and 9,958 liters (2,585 and 2,630 gallons) of water to produce 1 kg (2.2 lbs.) of cotton and that this represents 1.7 percent of the U.K.'s water footprint. N. Cherrett, J. Barrett, A. Clemett, M. Chadwick and M.J. Chadwick (2005), Ecological Footprint and Water Analysis of Cotton, Hemp and Polyester. Report prepared for and reviewed by Bioregional Development Group and World Wide Fund for Nature—Cymru. Stockholm Environment Institute, Stockholm. Available at www. organicexchange.org/Farm/Reading%20and%20References/Cotton%20 Hemp%20Polyester%20study%20SEI%20and%20Bioregional%20 and%20WWF%20Wales.pdf.

23. Impacts up to the farm gate are from Cranfield 2006. Impacts from the farm to the checkout are from Small World Consulting's work for Booths supermarkets. The main report is available online: "The Greenhouse Gas Footprint of Booths, July 2010," www.booths.co.uk/ Documents/Booths_Full_Report_100720.pdf.

24. See previous note.

25. There are studies giving lower figures than this, but I have also read suspicions that they underestimate. T. Garnett (2006), Fruit and

Vegetables & U.K. *Greenhouse Gas Emissions: Exploring the Relationship.* FCRN working paper 06-01. Food Climate Research Network (FCRN). Available at www.fcrn.org.uk/fcrnPublications/publications/PDFs/Fruitnveg_paper_2006.pdf.

26. Williams, Audsley and Sandars (2006) (see note 11 above). Although this looks like the best information around, it is contested. I know farmers who are highly critical of the assumptions made in the same report about organic dairy herds. My high-end figure is adjusted upwards from Cranfield's 38.6 kg CO_2e per kilo (38.6 lbs. per pound) to take account of produce from a colder time of year rather than the year-round average reported in the Cranfield study.

27. The figure for trout up to the slaughterhouse comes from a database of life-cycle analyses sponsored by the Danish government: P.H. Nielsen, A.M. Nielsen, B.P. Weidman, R. Dalgaard, and N. Halberg (2003), LCA Food Database, "Lifecycle assessment of basic food" (2000–2003). Aarhus University, Denmark, www.lcafood.dk. I have used Booths supermarkets' carbon impact model (see note 23 above) to estimate impacts from the slaughterhouse to the checkout.

28. See note 26 above. The figures for frozen filleted fish, fish products, unprocessed shellfish, and processed shellfish come from A. Wallen, N. Brant, and R. Wennersten (2004), "Does the Swedish consumer's choice of food influence greenhouse gas emissions?" *Envir Sci Policy* 7, 525–535. The final figure for cod comes from C. Foster et al. (2006), *Environmental Impacts of Food Production and Consumption. A Report to the Department for Environment, Food and Rural Affairs.* Manchester Business School. Defra, London, p. 101. Available at randd.defra.gov.uk/Document. aspx?Document=EV02007_4601_FRP.pdf. The agreement with the Danish figures above is encouraging. Unless otherwise stated, the figures come from Nielsen et al. (see note 27 above).

29. Figures from the Inventory of Carbon and Energy, a publicly available database of embodied energy figures for several hundred materials, compiled from the best-available life-cycle analyses around the world (G. Hammond and C. Jones [2008], *Inventory of Carbon and Energy (ICE),* version 1.6a. University of Bath).

30. N. Höhne, D. Phylipsen, and S. Moltmann (2007), *Factors Underpinning Future Action: 2007 Update.* A report by Ecofys for the Department for Environment, Food and Rural Affairs. Ecofys GmbH, Cologne. Available at www.fiacc.net/data/fufa2.pdf. The data here are extrapolated from Climate Fact Sheets for different nations.

10 kilos to 100 kilos (22 pounds to 220 pounds)

1. The widow of former Philippines president Ferdinand Marcos was listed by *Newsweek* as one of the 100 "greediest people of all time." She gained some of her notoriety from her shoe collection, gathered while plenty of her fellow citizens lived in poverty.

2. A weakness of the input–output model I used for this is that it assumes that Chinese production is as carbon efficient as U.K. manufacture. It isn't. It's worse. In reality, a key carbon decision for footwear suppliers is where to have product made.

3. U.S. Food Safety and Inspection Service, www.fsis.usda.gov/HELP/FAQs_Hotline_Illness/index.asp (accessed October 2009).

4. This is the additional footprint arising from your decision to make the commute given that everyone else is already on the road. It is also the difference you can make by stopping commuting. It is more than your fair share of the total pollution, which would only be double rather than three times the normal emissions from driving that distance on an empty road.

5. To make it very simple, think of a line 10 cars long, moving at one car per minute. Assuming the line has stayed the same size, those 10 cars will between them have lined up for 100 car minutes by the time they have all gone through. Add your car and you have 11 cars all lining up for 11 minutes. That's 21 minutes more lining up, even though you experience just 11 minutes. You get the same effect when you model slightly more complicated things such as ring roads with line ups at each roundabout. None of this takes account of the possibility that you are the person who gets stuck at the intersection, triggering gridlock and a whole new multiplier effect.

6. The Highway Code figures for typical stopping distances are 96 m/310 ft. (24 car lengths) at 70 miles per hour and just 53 m/174 ft. (13 car lengths) at 50 miles per hour. The stopping distance has two components: the thinking distance, which is proportional to your speed, and the larger braking distance, which is proportional to the square of your speed. On this basis a lane at 50 miles per hour can take nearly twice the traffic of one at 70 miles per hour. So there is no need for anyone to line up when the lane closes, provided that no one leaves it to the last moment to change lanes. In reality most drivers don't leave as much as their stopping distance between them and the car in front, but the principles here still apply if they keep leaving the same proportion of that stopping distance between themselves and the next car as they slow down.

7. The carbon footprint tool is available as a free download from the Cumbria Tourism website, at www.cumbriatourism.org/sustainable-tourism/carbon-footprint-toolkit.aspx. It sets out to include just about everything that a business buys and does. For the most part it uses the same input–output model that I have drawn upon in this book (see above note 4 in the "Under 10 grams" section. The calculator itself was put together on a limited budget but seems to work fine. It is designed for use by businesses of any size, and the ambitious intention is for this to be possible without businesses needing any external help. The tool was developed by my business, Small World Consulting. Thanks are due to Jessica Moss, who did most of the work, and to Cumbria Tourism and the Lake District National Park for funding assistance.

8. The data come mainly from G.P. Hammond and C.I. Jones (2008), *Inventory of Carbon and Energy (ICE) Version 1.6*. University of Bath. Available at www.organicexplorer.co.nz/site/organicexplore/files/ ICE%20Version%201.6a.pdf. See also E.M. Kalliala and P. Nousiainen (1999), "Life cycle assessment: environmental profile of cotton and poly-ester-cotton fabrics." *AUTEX Res J* 1 (1), 8-20. Available at www.autexrj. org/pdf/1999_No1/2.pdf.

9. Hammond and Jones (2008) (see note 8 above).

10. Estimates of the energy use per gigabit of transmission are 28 kg (62 lbs.) CO_2e over a UMST network and 31 kg (68 lbs.) CO_2e for GSM n. Based on M.F. Emmenegger, et al. (2004), Life Cycle Assessment of the mobile communication system UMTS: toward eco-efficient systems. *Int J Life Cycle Assessment* 11 (4), 265-276; and K. De Decker (2008), "The right to 35 mobiles," *Low-tech Magazine*, www.lowtechmagazine. com/2008/02/the-right-to-35.html.

11. Including transport to the store.

12. This is for a Nokia 7600, a fairly simple phone by today's standards. It has a camera and an MP3 player, but it doesn't do email or stop you from getting lost. Based on a Nokia life-cycle analysis summarized in a WEEE Man Case Study Snapshot (weeeman.org/html/what/lifecycle_case.html) and referenced in J. Quaiguasi Froto Neto (2008) "Eco-efficient supply chains for electrical and electronic products," PhD thesis, Erasmus University Rotterdam, available at publishing.eur.nl/ir/repub/asset/14785/ EPS2008152LIS9058921925Quariguasi.pdf. The figures are also broadly in line with those in M.L. Socolof, D. Cooper, and P. Dillon (2007), *Expansion of the Electronics Environmental Benefits Calculator: Mobile Phone Reuse and Recycling*. Report submitted to Eastern Research Group, Lexington, MA. Available at www.abtassociates.com/reports/eebc_ cellphone.pdf.

13. Compiled from data in the above four references.

14. The estimate comes from De Decker (2008) (see note 10 above). This is also a good source of links for anyone digging around to understand the carbon impact of cell phones and other communications technology.

15. Emmenegger, et al. (2004) (see note 10 above).

100 kilos (220 pounds) to 1 ton

1. A staggering 5 hours of life lost through death per 1,000 miles of driving. My sum was just this: loss of life expectancy per mile = 2,538 deaths on U.K. roads per year × 48 remaining years of life expectancy of an average driver, divided by 216 billion person car miles on U.K. roads per year = 5 hours life lost per 1,000 miles of driving. (National Travel Survey, Department of Transport, 2009). I've based my sums on your having a life expectancy of another 48 years (I picked a 40-year-old man with a healthy lifestyle and because it gives me a nice round number), but you

might want to adjust for your own situation. I haven't taken account of the fact that some of the deaths are of pedestrians (thinking that you might be just as bothered about killing others as you are yourself) but I also haven't taken into account the possibility that you might acquire one of the 26,000 serious injuries or 150,000 minor injuries that are served up to U.K. car users each year. It's a lot better to be injured than killed on the road, but injury happens 10 times more often. I have also assumed that highway journeys are on average safer per mile compared with other car trips.

2. AA Routeplanner, www.theaa.com/route-planner/index.jsp (accessed 4 March 2009).

3. The figure of 20 percent of presents being unwanted comes from G. Haq, A. Owen. E. Dawkins, and J. Barrett (2007) *The Carbon Cost of Christmas.* Stockholm Environment Institute. Available at 50plus. climatetalk.org.uk/downloads/CarbonCostofChristmas2007.pdf. The other numbers here are mine. This paper has an analysis in the same vein.

4. See answers.ask.com/Society/Holidays/ how_much_do_americans_spend_on_christmas.

5. In Monty Python's *Meaning of Life,* Mr. Creosote explodes in an unforgettable manner after being tempted into one last wafer-thin mint.

6. Energy Saving Trust is a body funded by the U.K. government that offers advice and grants: www.energysavingtrust.org.uk. Their sums are for a three-bedroom house and include reduced savings for "comfort uptake": turning up the temperature a bit when the new insulation is fitted and cashing in on the added comfort that becomes possible. I have adjusted their figures slightly by adding 10 percent to the carbon savings to take account of the emissions involved in supplying gas to your house as well as your burning it. I have also applied discount rates of 10 percent per year to the financial savings because this gives a more realistic picture (see Discount rates, page 131).

7. The ICE database gives a figure of 1.2 kg CO_2e per kilo (0.55 kg CO_2e per pound) for fiberglass insulation. I've gone with these despite the problems that process life-cycle analysis has with underestimating absolute numbers. G.P. Hammond and C.I. Jones (2008), *Inventory of Carbon and Energy (ICE) Version 1.6.* University of Bath. Available at www.organic-explorer.co.nz/site/organicexplore/files/ICE%20Version%201.6a.pdf. I have allowed a density of 18 kg per cubic meter (1.1 lbs. per cubic foot) and assumed an attic area of 65 square meters (700 sq. ft.). My figure probably undercooks it a bit, but the embodied emissions are still going to be a small deal in the overall sums.

8. I have assumed that the carbon footprint of extracting the gold has been "written off" by previous owners.

9. This seems a reasonable ballpark figure based on data in N. Höhne, D. Phylipsen, and S. Moltmann (2007), *Factors Underpinning Future*

Action: 2007 Update. A report by Ecofys for the Department for Environment, Food and Rural Affairs. Ecofys GmbH, Cologne. Available at www.fiacc.net/data/fufa2.pdf, in which, for example, greenhouse gas emissions per GDP for China are 2.3 times greater than those of the U.K. and carbon emissions per ton of steel are twice as high as in the U.K.

10. Energy Star, sponsored by the U.S. Environment Protection Agency, has downloadable spreadsheets showing typical energy consumption figures for all qualifying computers: www.energystar.gov (accessed October 2009).

11. Apple and the Environment, images.apple.com/environment/reports/docs/iMac_21_5_inch_Environmental_Report_2009.pdf (accessed February 2010).

12. Based on attributing the footprint of manufacture, transport, and facilities across their product range, with computers accounting for $14.3 billion of a $32.8 billion revenue. Sales data (2008) from Apple Watch, blogs.eweek.com/applewatch/content/corporate/apple_fiscal_2008_by_the_numbers.html (accessed October 2009).

13. The difference is also about what I would have expected. A report by Small World Consulting and the Crichton Carbon Centre, "The implications of truncation error in process-based lifecycle analyses of traditional buildings and their components" (Historic Scotland 2009), summarizes various academic studies from around the world, all concluding that in the construction industry, process-based life-cycle analysis typically succeeds in capturing only just over half of the total footprint of buildings. On the simple basis that computers are more complex than buildings, we might expect this "truncation error" to be even larger still in computer manufacture. In this way the data from Apple reaffirm the 560 kg (1,235 lbs.) result from the input-output analysis. (See above note 4 in the "Under 10 grams" section.)

14. K. De Decker (2009) "The monster footprint of digital technology," *Low-tech Magazine*, cites various life-cycle analysis studies. Technology changes fast. It's not clear whether the capability of a machine is going up faster than the efficiency of production. In 2002 a 2 g chip could hold 32 megabytes of memory and had an estimated footprint of 38 kilowatt-hours of energy.

15. It draws 37 watts when idle, 2 watts in sleep mode and less than 1 watt when turned off.

1 ton to 10 tons

1. Reported in the *Guardian*: "Online government reveals NHS price list" (6 February 2004), citing U.K. government figures.

2. 36.5p per kWh for new build. 41p for retrofit. See the U.K. government's Department for Energy and Climate Change website, www.decc.gov.uk/en/content/cms/news/pn10_010/pn10_010.aspx.

3. Chris Goodall, author of the excellent and acclaimed *How to Live a*

Low-Carbon Life (published by Earthscan in 2007; updated in 2010) also runs a very good website, Carbon Commentary (see, for example, www. carboncommentary.com/2009/07/15/686#more-686 [accessed October 2009]). He's done sums on the financial payback from micro-renewables. His numbers look at least as good as anyone else's.

4. I've taken account of the fading efficiency of the cells and also the probability that by the end of the 40 years even our dirtiest electricity will not be coming from a friendlier fossil fuel than coal.

5. The figures are derived by using Defra conversion factors and their suggested 1.9 emissions weighting factor. I have not added on their 9 percent uplift factor to take account of planes not taking the most direct route— so this is a "best route" scenario. The factors I have used do take into account fuel supply chains. This makes a difference of just a few percent. The embodied emissions in the plane and the footprint of airport infrastructure are not significant compared with the huge fuel burn of the jet engines. (So don't take airports too seriously if they tell you how carefully they are managing the carbon footprint of the airport building.) I reach similar figures running the model produced by David Parkinson and assuming a full flight. Overall this suggests that the numbers I am quoting are on the low side. The flight is unlikely to be full every time, and the Defra figures are optimistic for such long-range flights where so much of the take-off weight has to be fuel.

6. Here is a glimpse of the main issues: The amount of nitrous oxide that a jet engine produces varies with altitude. Its effect on ozone levels also depends on altitude. Furthermore the effect of that ozone on climate is altitude dependent. Planes also cause contrails under certain atmospheric conditions, and these are known to make a short-lived but large contribution to the greenhouse effect. The contrails themselves depend on temperature, weather conditions, the time of day, and, you've guessed it, altitude.

7. An emissions weighting factor of 1.9 can be inferred from the IPCC 4 assessment report, published by Cambridge University Press. This is also the figure suggested in Defra (2009) "Guidelines to Defra/DECC's GHG conversion factors for company reporting" (www.defra.gov.uk/ environment/business/reporting/pdf/20090928-guidelines-ghg-conversion-factors.pdf), Annex 6, footnote 10.

8. David J.C. MacKay's *Sustainable Energy without the Hot Air* (2009), published by UIT Cambridge Ltd, (www.withouthotair.com) neatly explains the physics of this and many other carbon questions, including the case for electric cars over the internal combustion engine.

9. David Parkinson (2006) "A new way forward for air traffic control" (Sensus Ltd.), www.sensus-dp.demon.co.uk.

10. At constant altitude, a plane needs to have lift equal to its own weight. It takes more energy to gain altitude, but that is turned into potential energy that can be regained in descent provided that traffic

control allows things to be done efficiently. David Parkinson of Sensus produced a model of fuel burn in flight that closely mirrors European Aviation Authority figures. It is great for looking at different scenarios of plane types and models over different distances and with different payloads. The most efficient distance for a 747 turns out to be somewhere around 3,000 nautical miles (about 5,600 km). On these short flights it is also possible to carry a substantial freight load, which is impossible on the longer-range flight because the plane would be too heavy to take off. If you are interested in David's model, contact him at newmodel@sensus-dp.demon.co.uk.

11. Ammonium nitrate (NH_4NO_3) fertilizer is 35 percent nitrogen by weight. The nitrous oxide (N_2O) that is released is 64 percent nitrogen by weight. The 1 percent of the nitrogen that is emitted is 0.55 percent of the original weight of the fertilizer in nitrous oxide, with a global warming potential 300 times that weight in CO_2 equivalent. So 1 to 5 percent nitrogen released to the atmosphere is 1.65 to 8.25 tons CO_2e per ton of fertilizer applied to the crop.

12. All the agricultural data in this section came from a lecture by Professor David Powlson during a visit to Lancaster University in November 2009. He is working with the Chinese government to get the message across to farmers.

10 tons to 100 tons

1. The death also causes a carbon savings, which I have not factored in.
2. See note 2 for the chapter above entitled "1 ton to 10 tons."
3. Rates vary for different renewables options. See the U.K. government's Department for Energy and Climate Change website, www.decc. gov.uk/en/content/cms/news/pn10_010/pn10_010.aspx.
4. A report by Sinclair Knight Merz estimates a £55 billion ($85 billion) investment to replace 40 percent of the U.K.'s electricity with renewable sources, mainly offshore wind, by 2020. That is 130 terawatt-hours per year (130,000 billion units per year). This pays back in 15 years even with a 10 percent discount rate and valuing generated electricity at just 5p (8 cents) per unit. Sinclair Knight Merz (2008) *Growth Scenarios for U.K. Renewables Generation and Implications for Future Developments and Operation of Electricity Networks* BERR Publication URN 08/1021, www.berr.gov.uk/files/file46772.pdf (accessed October 2009).
5. "A study of the embodied energy of upgrading or replacement options for traditional buildings." A report for Historic Scotland by Crichton Carbon Centre and Small World Consulting, October 2009. Available at www. historic-scotland.gov.uk/index/learning/freepublications.htm. The embodied emissions quoted here draw strongly upon input-output analysis (see above note 4 in the "Under 10 grams" section). Process-based approaches were also used, giving figures about 40 percent lower. This is

about as expected, given the systematic tendency for underestimation in process-based life-cycle analysis.

100 tons to 1 million tons

1. These calculations do not take account of carbon discount rates. In other words, I have worked from the principle that 1 ton of carbon emitted today has exactly the same impact as 1 ton of carbon emitted in the future. This is a broadly reasonable assumption if you take the view that future generations are just as important as our own and that the sensitivity of the planet to each additional ton of carbon will stay roughly the same throughout your child's life.

2. Optimum Population Trust, "Contraception is 'greenest' technology" (9 September 2009), www.optimumpopulation.org/releases/opt. release09Sep09.htm.

3. The original sum was £25 million, equivalent to about $37 million at the time of writing. The figures come from a study I was involved in for a pool in a town in Scotland in 2007. I can't say which because, even though I'm sure they wouldn't mind being named, I haven't asked them whether they would mind.

4. Worldwatch Institute (2009), *State of the World 2009*, 26th ed. (Earthscan, London), p. 32.

5. Mongabay.com, www.mongabay.com (accessed October 2009).

6. Amazon Fund, www.amazonfund.org.

7. J. Tollefson (2009), "Paying to save the rainforests." *Nature* 460, 936–937.

8. Rhett A. Butler, mongabay.com (November 16, 2005). "World deforestation rates and forest cover statistics, 2000–2005." Available at news.mongabay.com/2005/1115-forests.html. All data derived from the Forest Resources Assessment and the State of the World's Forests published by the UN Food and Agriculture Organization (FAO).

9. Shuttle data from Wikipedia. Other figures in my calculations were: 31 MJ per kg (14 MJ per lb.) for the solid fuel, 143 MJ per kg (65 MJ per lb.) for the hydrogen. I used 70 g CO_2e per MJ as a general figure for emissions from the burning of fossil fuels and added 10 percent for their supply chains up to the point of combustion.

10. "What do you care what other people think?" (Richard Feynman, 1989) is a fascinating and entertaining account of the technical and management failures behind the disaster. Also recommended for anyone who is trying to get some clear thinking into a bureaucracy.

11. 4.6 tons per return trip if you live in Hong Kong. Flying London to Hong Kong is about the same as Los Angeles to Barcelona, see page 133.

1 million tons and beyond

1. This estimate comes from the British Geological Survey (2005). The U.S. Geological Survey estimates just 200 million tons.

2. David McCandless and Ben Bartels (2010), "Planes or volcano?", *Information Is Beautiful*. Available at www.informationisbeautiful. net/2010/planes-or-volcano (accessed July 2010).

3. If you've read the book from the start you will have gathered already that this list is just the easy bits and you could happily double the footprint if you were a bit more inclusive. It's best not to get too bothered on this occasion. The numbers come from *Feasibility Study for a Carbon Neutral 2010 FIFA World Cup in South Africa*, Department of Environmental Affairs and Tourism, Republic of South Africa, and Norwegian Embassy, 2009. Available at www.norway.org.za/NR/rdonlyres/3E6BB1 B1FD2743E58F5B0BEFBAE7D958/114457/FeasibilityStudyforaCarbon-Neutral2010FIFAWorldCup.pdf.

4. Ignoring the detail that the World Cup only goes on for a few weeks and isn't even on all the time during that period.

.5. The numbers are derived from Gartner estimates and U.K. Market Transformation Programme reports: *Case Study: EU Code of Conduct for Data Centres: Reducing the Energy Consumed by BT Data Centres (BT/Defra Pilot)* (February 2009) and *Global Carbon Impacts of Energy Using Products: Report for Defra/the Market Transformation Programme by Klinckenberg Consultants* (April 2009). Available from www.mtprog. com (accessed October 2009). Various sources gave a picture consistent with my numbers, including peakenergy.blogspot.com/2008/09/ cutting-data-centre-energy-demand.html and news.cnet.com/Gartner-urges-action-on-data-center-emissions/2100-1022_3-6212965.html.

6. *The Rebound Effect: An Assessment of the Evidence for Economy-wide Energy Savings from Improved Energy Efficiency* (October 2007), U.K. Energy Research Centre. Available at www.ukerc.ac.uk/Downloads/ PDF/07/0710ReboundEffect/0710ReboundEffectReport.pdf.

7. The figures I used are at the low end of the estimates given. The 2009 Australian forest fires covered at least 450,000 hectares (Wikipedia, 2009). Asa Wahlquist, in *The Australian* (13 February 2009), quotes Mark Adams from the University of Sydney in reporting that the area affected contained over 100 tons of carbon per hectare. Carbon forms 3.67 times its weight in CO_2.

8. This figure does not include an emissions weighting multiplier to take account of the additional impact of high-altitude aviation emissions (A report by Ecofys for the U.K.'s Department for Environment, Food and Rural Affairs in 2007, "Factors Underpinning Future Action. 2007 Update")

9. Worldwatch Institute (2009) *State of the World 2009*, 26th ed. (Earthscan, London).

10. Office of National Statistics (2009), "U.K. environmental accounts: total greenhouse gas emissions by 93 economic sectors 1990 to 2007," available at www.statistics.gov.uk/statbase/ssdataset.asp?vlnk=5695.

This model underestimates the emissions from overseas because it is based on the assumption that overseas production is exactly as carbon intensive as the U.K. equivalent. We know this is flawed, because very often overseas industries are less energy efficient and also the electricity supply is often more carbon intensive. China is an obvious example where this is true.

Also included here in this analysis is 1.9 markup factor for aviation emissions to take account of the higher impact that high-altitude emissions are known to have. The pie also includes "fixed capital formation": new buildings and other new infrastructure that although not "consumed" are nevertheless something we continually demand.

11. E. Audsley, M. Brander, J. Chatterton, D. Murphy-Bokern, C. Webster, and A. Williams (2010), *How Low Can We Go? An Assessment of Greenhouse Gas Emissions from U.K. Food System and the Scope for Reduction by 2050*. WWF-UK.

12. This includes an emissions weighting factor of 1.9 for high-altitude emissions, as I explain on page 7.

13. N. Höhne, D. Phylipsen, and S. Moltmann (2007), *Factors Underpinning Future Action: 2007 Update*. A report by Ecofys for the Department for Environment, Food and Rural Affairs. Ecofys GmbH, Cologne. Available at www.fiacc.net/data/fufa2.pdf, "Country fact files." The figures do not include international aviation and shipping. For the U.K. that would add about 10 percent.

14. T. Jackson (2009), *Prosperity without Growth: Economics for a Finite Planet*. Earthscan, London. A recommended read.

15. Kilotons of TNT equivalent.

16. Duncan Clark in www.guardian.co.uk, "The carbon footprint of nuclear war" (2 January 2009), drawn from M.Z. Jacobson (2009), "Review of solutions to global warming, air pollution, and energy security," *Energy Envir Sci* 2, 148-173, DOI:10.1039/b809990c (first published as an Advance Article on the web on 1 December 2008: www.stanford.edu/group/efmh/jacobson/PDF%20files/ReviewSolGW09.pdf).

17. *The Three Trillion Dollar War* by Joseph Stiglitz, a Colombia University professor who won the Nobel Prize for Economics in 2001, and Linda Bilmes. See www.democracynow.org/2008/2/29/exclusive_the_three_trillion_dollar_war.

18. Using conversion factors of 0.33 and 0.22 kg (0.73 and 0.49 lbs.) CO_2e per pound sterling for U.K. output of defense and health services, respectively, and $1.6 per pound sterling averaged over the duration or the war so far. The figures assume that the U.S. and U.K. industries have the same carbon intensity. The model excludes direct emissions from combat itself. A large margin for error has been added in.

19. Worldwatch Institute (2009), *State of the World 2009*, 26th ed. (Earthscan, London): Chapter 6, pp. 56-58, "Reducing black carbon." Everything on black carbon is taken from this chapter.

20. Intergovernmental Panel on Climate Change (2007), *IPCC Fourth Assessment Report. Working Group I Report: The Physical Science Basis* (Cambridge University Press, Cambridge, Chapter 2; Ramanathan & Carmichael, op. cit. note 2.) Referenced in the Worldwatch Institute's piece on black carbon (see note 9 above). Radiative forcing from black carbon is put at 0.4 to 0.9 watts per square meter (0.04 to 0.08 watts per square foot), in contrast with 1.6 watts per square meter (0.15 watts per square foot) for CO_2.

21. Intergovernmental Panel on Climate Change (2007), Global anthropogenic GHG emissions. In *Climate Change 2007: Synthesis Report* (IPPC, Geneva), p. 36. Adapted to include an emissions weighting factor of 1.9 for high-altitude emissions.

22. © SASI Group, University of Sheffield.

23. © SASI Group, University of Sheffield.

24. © SASI Group, University of Sheffield.

25. Wikipedia list of oil spills, available at en.wikipedia.org/wiki/ List_of_oil_spills (accessed 25 July 2010).

26. Earthtrends. World Resources Institute, Searchable Database, 2006 data. Available at earthtrends.wri.org/searchable_db (accessed October 2009).

More about food

1. Most of the figures here come from the input–output model (see above note 4 in the "Under 10 grams" section) and are in line with other estimates. Tim Jackson's paper for the Carbon Trust, "The carbon emissions generated in all that we consume" (January 2006), www.carbontrust. co.uk/Publications/pages/publicationdetail.aspx?id=CTC603, includes estimates of the CO_2 emissions from cooking and storing food at home. See also "Cooking up a storm" by Tara Garnett, Food Climate Research Network, University of Surrey (2008), www.fcrn.org.uk/ fcrnPublications/publications/PDFs/cuas_Summary_web.pdf.

2. Agricultural greenhouse gas emissions are from U.K. environmental accounts Office of National Statistics (2009) "U.K. environmental accounts: total greenhouse gas emissions by 93 economic sectors 1990 to 2007," www.statistics.gov.uk/statbase/Expodata/Spreadsheets/D5695. xls (accessed 25 January 2008).

3. The proportions for the U.S. are similar: 2010 U.S. Greenhouse Gas Inventory Report, available at www.epa.gov/climatechange/emissions/ usinventoryreport.html3. *Guidelines to Defra/DECC's GHG Conversion Factors for Company Reporting* (2009). Produced by AEA for the Department of Energy and Climate Change (DECC) and the Department for Environment, Food and Rural Affairs (Defra).

4. See for example, Steven C. Slater, and Tillman U. Gerngross (August 2000), "How Green Are Green Plastics?" *Scientific American*, available at

www.scientificamerican.com/article.cfm?id=how-green-are-green-plast, and "Degradable, Biodegradable, Compostable," *Green Living Tips*, available at www.greenlivingtips.com/articles/197/1/Degradable-Biodegradable-Compostable.html.

5. From WRAP (2008), *The Food We Waste*. Waste & Resources Action Programme (WRAP), Banbury. Available at www.wrap.org.uk/downloads/The_Food_We_Waste_v2__2_.ec417f6e.5635.pdf. The overall figure of 30 percent waste includes bones and other bits that we don't all consider edible.

6. These include CO_2, which has a global warming potential (GWP) of 1, of course, but requires a high-pressure system.

7. See note 5 above.

Further information

1. Howard Friel's book *The Lomborg Deception* (Yale University Press, 2010) does a thorough analysis of the flaws in Lomborg's work.

2. The two I'm most often asked about are Nigel Lawson's *An Appeal to Reason: A Cool Look at Global Warming* (Duckworth, London, 2008) and Channel 4's mischievous 2007 documentary *The Great Global Warming Swindle*.

3. J. Minx, T. Wiedmann, J. Barrett, and S. Suh (2008), *Methods Review to Support the PAS for the Calculation of the Embodied Greenhouse Gas Emissions of Goods and Services*. A research report for the Department for Environment, Food and Rural Affairs by the Stockholm Environment Institute and the University of Minnesota. Defra, London. Available at randd.defra.gov.uk/Document.aspx?Document=EV02074_7071_FRP.pdf.

Acknowledgments

MY BIGGEST THANKS go to Liz, Bill, and Rosie for brilliant support and understanding, especially over Christmas when the deadlines loomed largest.

The book could not have happened without Duncan Clark, Mark Ellingham, and others at Profile. Duncan's edits and advice have been superb throughout the project.

Many thanks to Jess Moss at Small World for unearthing quirky data and sorting out dozens of references—and for reminding me to get a move on—and to Helen Cammack and Claire Hoolohan for help with the tricky job of revising the book for a North American audience.

David Howard, Kim Kaivanto, Andy Scott, and Geraint Johnes from Lancaster University, and Sonny Khan, all helped with the input-output model that I have drawn upon extensively. Thanks also to David Parkinson and Chris Goodall, among others, for answering technical queries.

Andrew Meikle let me chatter away during ride shares and has been a frequent sounding board. He read early pages aloud so that I could hear how bad they were. Others who cast a friendly eye include Phil and Jane Latham, Aly Purcell, Rachel Nunn, and Mark Jameson. Mum and Dad, true to form as incredible parents, both picked through the entire draft at a moment's notice.

Thanks to Kim Quazi for helping me thrash out the first ideas years ago in a pub and to Kirsty Pelling, the journalist in the supermarket. If I go even further back, the list of people to be grateful to is clearly endless, but I want to thank Lee Pascal, David Brazier, and Simon Loveday for three very different contributions.

I'm grateful to many of Small World's clients for providing material but especially to Booths supermarkets, Lancaster University, the Crichton Carbon Centre, Historic Scotland, and the Keswick Brewing Company.

Thanks to everyone who said "Oh, you're writing a book... how interesting!" and to those who, just to keep my morale up, preordered copies long before I'd even finished the first draft.

Finally, thanks are also due to Richard Hampton, Hans Ohanian, Jean Sireyjol, and others who, since the first print run, have suggested a variety of improvements that are now incorporated into the text.

Index

Endnotes are indicated by "n" followed by the note number.

hamburgers. *See* meat
hands, drying of, 21–22
Harrison, John, 76
heart bypass surgery, 129–30
heat, units of, 54–56
high-altitude emissions. *See* aviation
hospitality industry, 108–11, 163
hothouses, 41, 85, 99–100, 178, 181
houses. *See* construction industry
human beings: individual
    footprints, 11, 137–38, 149–50,
    161, 169; value of, 13–15.
    See also mortality
hydrogen generation, 153

ice cream, 54
Iceland, 54, 56–58, 158, 162,
    165–66, 167, 194
India, 103, 104, 152, 162, 165–66,
    173, 194
indirect emissions, defined, 7
Indonesia, 152, 194
input–output analysis, 124, 141–42,
    191–93
insulation, 120–22, 188, 189,
    215nn6–7
Intergovernmental Panel on Climate
    Change (IPCC), 185–86
Inventory of Carbon and Energy, 191
investment, in renewable energy,
    59–61
Iran, 194
Iraq wars, 168, 174
ironing, 25–26

jewelry, 122–23
junk mail. *See* mail

Kemp, Roger, 43
Kenya, 208n35
Keswick Brewing Company, 50–51

lamb, 111. *See also* meat
Lancaster University, 154–56
landfills. *See* waste management
La Paz (Bolivia), 39
laundry, 25–26, 39–41, 86–87,
    95–96, 209nn8–9

Leontief, Wassily, 192
life-cycle analysis, 124, 141–42,
    190–91, 192
lights and light bulbs, 102
Lomborg, Bjørn, 184–85
London (U.K.), 44, 45

mail, 21, 45–47, 114, 204–5n11,
    205n12
Malawi, 11, 137, 162, 167
Marcos, Imelda, 105, 212n1
marginal demand, 44, 57, 206n17
Market Transformation
    Programme, 191
measurement, units of, 10–11
meat: beef and burgers, 12, 22, 26,
    42, 88–89, 97, 112, 158; lamb,
    111; large footprint of, 177–78;
    reducing consumption, 180–81
metals, 61–63, 103–4, 122–23
methane: from agriculture, 91–92,
    176–77; from egg production, 98;
    as greenhouse gas, 5, 6, 196n2,
    196n3; from livestock, 73, 75,
    88–89, 97, 111; world totals, 170,
    171, 172, 173
Miliband, Ed, 40–41
milk. *See* dairy products
mortality: from car travel, 214–15n1;
    from climate change, 13–14, 15,
    197–99n6; from war, 169
mortgages, 126–28

Netherlands, 70, 195, 208n35
newspapers, 48–50
New Zealand, 30, 104, 195
Nigeria, 103, 104, 195
nitrous oxide: from agriculture and
    fertilizer, 97, 136–37, 176–77,
    218n11; from aviation, 217n6;
    from egg production, 98; as
    greenhouse gas, 5–6, 196n2;
    world totals, 170–71, 172, 173
North America, 11, 16, 96, 110,
    137–38, 149–50. *See also* Canada;
    United States
Norway, 57, 104, 153, 195

nuclear electricity, 43–44, 56, 57, 64
nylon, 95

oil spills, 174
oranges, 32–33, 99

packaging: bottled water, 44–45;
    of eggs, 98; of fish, 101; food and
    beverages, 181; and food wastage,
    41; recycling of, 178–79; reduc-
    tion of, 181; of wine, 80–81.
    See also plastic
pants, 94–97
paper: in books, 77–79; and digital
    information, compared, 160–61;
    in letters and junk mail, 45–47;
    newspapers, 48–50; paper bags,
    24–25; toilet paper, 65–66, 153;
    in U.K.'s national footprint, 163,
    208–9n1
payback periods, 120, 121, 131–32,
    145–47, 188
Peru, 85, 152
Philippines, 16
photovoltaic panels. *See* solar hot
    water and electricity
Pinatubo, Mount, 157
plastic, 44, 82–83, 178–79, 191,
    204n7
plastic bags, 22–23, 24, 133, 202n7
porridge, 52–53, 103, 111. *See also*
    breakfast cereal
postal service. *See* mail
potatoes, 71–72
power meters, 36
pricing issues, of emissions
    reduction, 14–15
process life-cycle analysis, 124,
    141–42, 190–91, 192

rebound effect, 21, 161
recycling and reuse: of beer contain-
    ers, 52; and disposal as trash,
    compared, 61–64, 207n24; of
    glass, 80–81; of packaging, 178–
    79; of paper, 22, 24–25, 45–47,
    48–50, 65–66, 77–78;

of plastic, 23, 82–83; of steel, 103;
    of TVs, 34–35. *See also* waste
    management
Rees, Martin, 196n1
refrigerant gases, 6, 171, 179–80,
    196n2
refrigeration, 33, 54, 74, 100–101,
    179–80
renewable energy, 57–60, 218n4.
    See also specific technologies
restaurants. *See* hospitality industry
rice, 91–93, 210n15
roses, 70–71
Rumensin, 75
Russian Federation, 104, 152, 162,
    165–66, 167, 195
Rwanda, plastic bag ban, 22

SeaChoice purchase guide, 100
Seawater Greenhouse, 94
sheep, 111
shoes, 105–6
showers, 53–54. *See also* baths
Shumacher, Shelly, 47
solar hot water and electricity,
    54–55, 60, 83, 93–94,
    130–33, 189
South Africa, 104, 158–60, 195
space flight, 153–54, 158.
    See also aviation
space heating, 19–20, 54–56,
    120–22, 200–201n9, 205–6n15
Spain, 94, 195
spam (email), 20–21
standby mode, 35–36
steel industry, 103–4
stoves, 27–29. *See also* cooking
strawberries, 41
sustainable lifestyle, 77
Sweden, 104, 195
swimming pools, 150–51
Sydney (Australia), desalination, 93

taxes, on email, 21
tea, 37–38
television, 33–36, 158–59,
    202–3n11, 203n12

And finally... this period:

•

It's a particularly large period. I estimate that at 2 microns thick and 1 millimeter wide, it weighs about one five-hundredth of a milligram. At perhaps 10 kg per kilo (4.5 kg CO_2e per pound) for the ink that's a footprint of one-fiftieth of a milligram of CO_2e.

If just a few readers of this book have spent just a few seconds in quiet, low-carbon, contemplation of this black dot, then it will have paid back its impact many more times over than the world's best offshore wind farms can ever hope to achieve. If it has distracted just a few people for just a few seconds from their shopping sprees, ski holidays, car journeys, and Peruvian asparagus, then, for its size, it will have made a truly outstanding contribution to the low-carbon world. An inspiration to us all.